Notes on Medical Bacteriology

EDITED BY

J. Douglas Sleigh
Senior Lecturer in Bacteriology

Morag C. Timbury
Titular Professor and William Teacher Lecturer in Bacteriology

The Department of Bacteriology,
University of Glasgow,
Royal Infirmary, Glasgow

FOREWORD BY

Sir James W. Howie
Formerly Director, Public Health Laboratory Service

CHURCHILL LIVINGSTONE
EDINBURGH LONDON MELBOURNE AND NEW YORK 1981

CHURCHILL LIVINGSTONE
Medical Division of Longman Group Limited

Distributed in the United States of America by Churchill
Livingstone Inc., 1560 Broadway, New York, N.Y.
10036, and by associated companies, branches and
representatives throughout the world.

First published 1981
Reprinted 1983
Reprinted 1984

ISBN 0 443 02264 X

British Library Cataloguing in Publication Data
Notes on medical bacteriology. — (Churchill
Livingstone medical text)
1. Bacteriology, Medicine
1. Sleigh, J. Douglas
II. Timbury, Morag C.
616'.014   QR46

Library of Congress Catalog Card Number 81-67931

Printed in Hong Kong by
Wing King Tong Co Ltd

Foreword

These admirable notes represent an entirely successful effort to present the main facts of medical microbiology in the fewest possible words. They will surely come as a boon and a blessing to all students who require an introduction to the subject which picks out the information that matters most from the now embarrassingly large volume of good published work. This mass of riches makes the work of authors of supposedly elementary textbooks very difficult indeed. The essence of art is selection, but the selection must be properly balanced, and therein lies the difficulty. In teaching medical microbiology to medical undergraduates, to young medical graduates who are beginning to specialise in the subject, and to science graduates and scientific officers who are to work in medical laboratories, it is important to give as an introduction enough about the biology of the microbes to make clear how they may be identified so that the correct specimens are sent for examination, the correct methods used to that end, and the correct interpretations put upon the laboratory findings. Matters of detail and method soon become second nature to experienced microbiologists in their day-to-day work; but the new student must not be put off the subject by thinking that he must memorise from the beginning all the details in the cookery book of the microbiologists' kitchen. What these notes do so well is to set out the necessary basic facts and principles clearly and briefly for noting and easy reference, and then to concentrate on the exciting facts about where microbes live, how they get around, how they may be contained, what diseases they produce, and how these diseases may be prevented or treated. The concerns both of individual patients and of the community are kept in proportion. Immunisation, hygiene, epidemiology, and antimicrobial therapy are all competently dealt with alongside the facts about the microbes concerned. The result is a miniature that is also a masterpiece. I have pleasure in writing this foreword and in wishing the book the success I know it deserves.

J. W. H.

Preface

This book is intended primarily for medical students studying for the Professional Examination in Microbiology, but we hope it may also be useful for dental and nursing undergraduates. It has been written as a collaborative effort by the teaching staff of the Department of Bacteriology in the Royal Infirmary, Glasgow and aims to give a concise account of medical bacteriology. Systematic bacteriology is dealt with briefly and we have tried to emphasise the clinical aspects of the subject which seem to us of most importance in present-day medical practice. It should be supplemented by reading from a larger textbook and some recommended books for further reading are listed on page 344.

We thank many colleagues for advice and much useful discussion, notably Mr R. Brown, Professor J. G. Collee, Dr Heather M. Dick, Dr R. J. Fallon, Dr Eve Kirkwood, Dr M. Laidlaw, Mrs Janet McCabe, Dr G. Masterton, Dr D. Reid and Dr Jean Thomson. Professor K. Hodgkin allowed us to quote from his excellent book *Towards Earlier Diagnosis in Primary Care*. Thanks are also due to Professor J. Hume Adams, Dr R. St C. Barnetson, Professor W. Brumfitt, Dr R. S. Kennedy, Dr A. Lyell and Dr P. S. MacFarlane, who took a great deal of trouble to provide us with slides and photographs. We are grateful to Mr W. McCormick, for allowing us to reproduce some of the excellent slides he has prepared for the department over many years. We thank Mrs Alison Fletcher for typing the manuscript and Mrs Brenda Burns and Mr Ian McKie for preparing the drawings and photographs respectively. We would also express our thanks to Dr Valerie Inglis for compiling the index.

Both of us owe a particular debt to Sir James Howie not only for writing the foreword to the book but for our early years of training in his department. He inspired a whole generation of medical microbiologists.

Glasgow, 1981

J. Douglas Sleigh
Morag C. Timbury

Contributors

Dr C. G. Gemmell, B.Sc., Ph.D. M.I.Biol.
Dr R. Hardie, M.B.Ch.B., B.Sc., M.R.C.Path.
Dr G. R. Jones, B.Sc., Ph.D.
Dr T. W. MacFarlane, D.D.S., M.R.C.Path.
Dr J. McGavigan, M.B.Ch.B., M.R.C.Path.
Dr D. J. Platt, B.Sc., Ph.D.
Dr J. D. Sleigh, M.B.Ch.B., M.R.C.P.(Glasg.), F.R.C.Path.
Dr G. Sweeney, B.Sc., Ph.D.
Professor M. C. Timbury, M.D., Ph.D., F.R.S.E., F.R.C.P. (Glasg.), F.R.C.Path.
Dr P. A. Wright, M.B.Ch.B., M.R.C.Path.

Diagrams and Drawings
Dr D. J. Platt

Contents

SECTION I: BACTERIAL BIOLOGY
1. Introduction   3
2. Bacteria: organisation, structure, taxonomy   6
3. Growth and nutrition of bacteria   14
4. Bacterial genetics   22
5. Laboratory methods   33
6. Sterilisation and disinfection   47

SECTION II: MEDICALLY IMPORTANT BACTERIA
7. Staphylococcus and Micrococcus   57
8. Streptococcus and Pneumococcus   62
9. Enterobacteria   71
10. Pseudomonas and other aerobic Gram-negative bacilli   80
11. Vibrio and Campylobacter   85
12. Parvobacteria   90
13. Corynebacterium and related bacteria   97
14. Mycobacterium   103
15. Actinomyces and Nocardia   106
16. Neisseria   109
17. Bacillus   113
18. Clostridium   115
19. Bacteroides and other non-sporing anaerobes   120
20. Lactobacillus   124
21. Legionella   125
22. Spirochaetes   127
23. Mycoplasma   131
24. Candida   133

SECTION III: BACTERIAL DISEASE
25. Normal flora   139
26. Host–parasite relationship   143
27. Epidemiology   151

28.	Specimens for bacteriological investigation	158
29.	Respiratory tract infections	161
30.	Diarrhoeal diseases	179
31.	Botulism	198
32.	Enteric fever	200
33.	Urinary tract infections	205
34.	Meningitis	211
35.	Sepsis	217
36.	Pyrexia of unknown origin	234
37.	Septicaemia and endocarditis	238
38.	Tuberculosis and leprosy	247
39.	Infections of bone and joint	258
40.	Sexually transmitted diseases	262
41.	Infections of the eye	271
42.	Oral and dental infections	275
43.	Zoonoses	291
44.	Infection in compromised patients	304
45.	Infections in general practice	308

SECTION IV: TREATMENT AND PREVENTION OF
BACTERIAL DISEASE
| 46. | Antimicrobial therapy | 313 |
| 47. | Prophylactic immunisation | 329 |

| | Recommended reading | 344 |
| | Index | 345 |

Bacterial biology

1

Introduction

Medical students need to learn bacteriology in order to diagnose and treat bacterial infections successfully.

Bacterial disease is still widespread and common: but its spectrum is changing so that once familiar diseases are now rare. Increasingly, the work of bacteriology laboratories is concerned with infection in patients in general hospitals — as distinct from fever hospitals — and in general practice.

The following are among the most important aspects of bacteriology which doctors must know in order to deal with infection.

Pathogenesis. The ways in which bacteria produce disease in the human body — essential information for diagnosis and treatment.

Diagnosis. Laboratory investigation depends on taking correct specimens and being able to assess the results obtained from the laboratory.

Treatment. Bacterial disease is one of the few conditions in medicine for which specific and highly effective therapy is available.

Epidemiology. The spread, distribution and prevalence of infection in the community.

Prevention. Many bacterial diseases have been virtually eradicated by immunization, public health measures and improved living standards.

HISTORY

Contagion. Since biblical times it has been known that some diseases spread from person to person.

The following are some of the pioneers responsible for the science of bacteriology as it is today.

Antony Van Leeuwenhook: a Dutch draper who made a microscope and in 1675 observed 'animalcules' in samples of water, soil and human material.

3

Louis Pasteur, the founder of modern microbiology: over a long period of brilliant and active research from 1860 to 1890, he developed methods of culture and showed that microorganisms cause disease. He also established the principles of immunisation.

Joseph Lister was Professor of Surgery in Glasgow Royal Infirmary. He applied Pasteur's observations to the prevention of wound sepsis — then almost an inevitable and often fatal complication of surgery. He discovered in 1867 an antiseptic technique to kill bacteria in wounds and in the air with carbolic acid. This revolutionised surgery.

Robert Koch was a German general practitioner who discovered the bacterial causes of many diseases — including tuberculosis in 1882. The discovery of agar as a setting agent for bacteriological media is attributed to Frau Koch from observations made in her kitchen. Koch defined the criteria for attributing an organism as the cause of a specific disease. These are the famous **Koch's postulates** and are as important today as when he propounded them:

1. The organism is found in all cases of the disease and its distribution in the body corresponds to that of the lesions observed.
2. The organism should be cultured outside the body in pure culture for several generations.
3. The organism should reproduce the disease in other susceptible animals.
   Nowadays a fourth postulate would be added:
4. Antibody to the organism should develop during the course of the disease.

Note. Many infectious diseases of which the cause is clearly identified do not fulfil the third and even occasionally the second of Koch's postulates.

Immunisation
The first successful immunisation was the demonstration by *Edward Jenner* in 1796 that a related but mild virus disease — cowpox — gave protection against subsequent attack by smallpox. Later, Pasteur's observations led to the development of the vaccines now widely and successfully used in medicine against diseases, e.g. diphtheria, tetanus, poliomyelitis, etc.

Antibiotics
The discovery of penicillin in 1929 by *Alexander Fleming* — a

Scot from Ayrshire — ushered in the antibiotic era. Generally derived from soil microorganisms, antibiotics kill or inhibit a wide variety of bacteria without harming their human host.

Public health

The development of a safe water supply, disposal of sewage, good housing and nutrition have also contributed largely to the decline in epidemic infectious disease.

2

Bacteria: organisation, structure, taxonomy

Bacteria are a heterogeneous group of unicellular organisms: their cellular organisation is described as *prokaryotic* (i.e. having a primitive nucleus) and differs from that of *eukaryotes* (plants, animals). Some of the main differences are listed in Table 2.1.

Table 2.1 Differences between prokaryotic and eukaryotic cells

Property	Prokaryotic cells	Eukaryotic cells
Chromosome number	One	Multiple
Nuclear membrane	Absent	Present
Mitochondria	Absent	Present

Genome. The most fundamental difference between bacteria and eukaryotes is that the bacterial chromosome, or genome, is a single circular molecule of double-stranded DNA; there is no nuclear membrane. Bacteria may from time to time harbour other smaller circular DNA molecules or plasmids which code for certain non-essential functions.

STRUCTURE

Bacteria have a rigid wall which determines their shape — only plants amongst other forms of living organisms possess this.

Shape. Bacteria may be:
1. Spherical — cocci
2. Cylindrical — bacilli
3. Helical — spirochaetes

Arrangement depends on the plane of successive cell divisions: e.g. chains — streptococci; clusters — staphylococci; diplococci — pneumococci; angled pairs or palisades — corynebacteria.

Gram's stain divides bacteria into Gram-positive or Gram-negative, an important step in classification. The Gram-staining reaction depends on the structure of the cell wall.

A diagram of a typical but composite bacterium is shown in Figure 2.1. Bacteria are cells with a rigid cell wall which surrounds the protoplast; this consists of a cytoplasmic membrane enclosing internal components and structures such as ribosomes and the bacterial chromosome.

External structures

External structures which protrude from the cell into the environment are present in many bacteria. These structures are:

1. *Flagella:* elongated filaments responsible for motility; composed of a protein 'flagellin' — chemically similar to myosin (Fig. 2.2).

2. *Pili:* finer shorter filaments extruding from the cytoplasmic membrane; also protein (pilin), they are responsible for adhesion (common pili) and probably for conjugation when genes are transferred from one bacterium to another (sex pili).

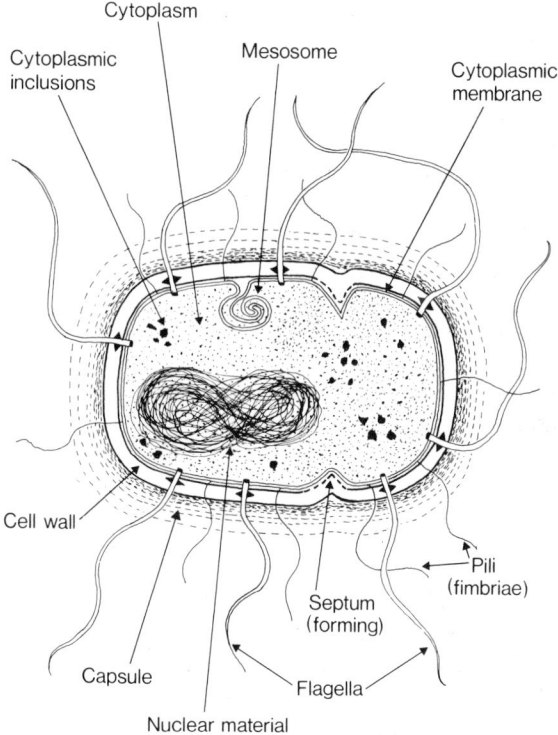

Fig. 2.1 Diagram of a typical but composite bacterial cell.

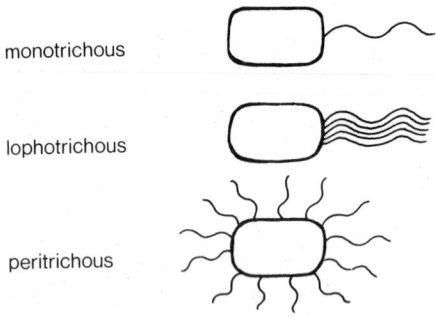

monotrichous

lophotrichous

peritrichous

Fig. 2.2   Distribution of flagella on bacteria.

3. *Capsules:* amorphous material which surrounds many bacterial species as their outermost layer; usually polysaccharide, occasionally protein; often inhibit phagocytosis and so their presence correlates with virulence in certain bacteria.

Cell wall

In addition to conferring rigidity upon bacteria the cell wall also protects against osmotic damage. It is porous and permeable to substances of low molecular weight.

Structure of the cell wall differs in Gram-positive and Gram-negative bacteria; this is illustrated in Figure 2.3.

Chemically the cell wall is peptidoglycan: this is a mucopeptide composed of alternating strands of *N*-acetyl muramic acid and *N*-acetyl glucosamine cross-linked with peptide subunits (Fig. 2.4).

Teichoic or teichuronic acids are part of the cell wall of Gram-positive bacteria: they maintain the level of divalent cations outside the cytoplasmic membrane.

Other components which may be present in the cell wall are antigens such as the polysaccharide (Lancefield) and protein (Griffith) antigens of streptococci and the lipopolysaccharide O antigens of Gram-negative bacilli.

Bacteria with defective cell walls

Bacteria develop and can survive with defective cell walls: these can be induced by growth in the presence of antibiotics and a hyperosmotic environment to prevent lysis.

Bacteria without cell walls are of four types:

1. *Protoplasts:* derived from Gram-positive bacteria and totally lacking cell walls; unstable and osmotically fragile; produced artificially by lysozyme and hyperosmotic medium: require hyperosmotic conditions for maintenance.

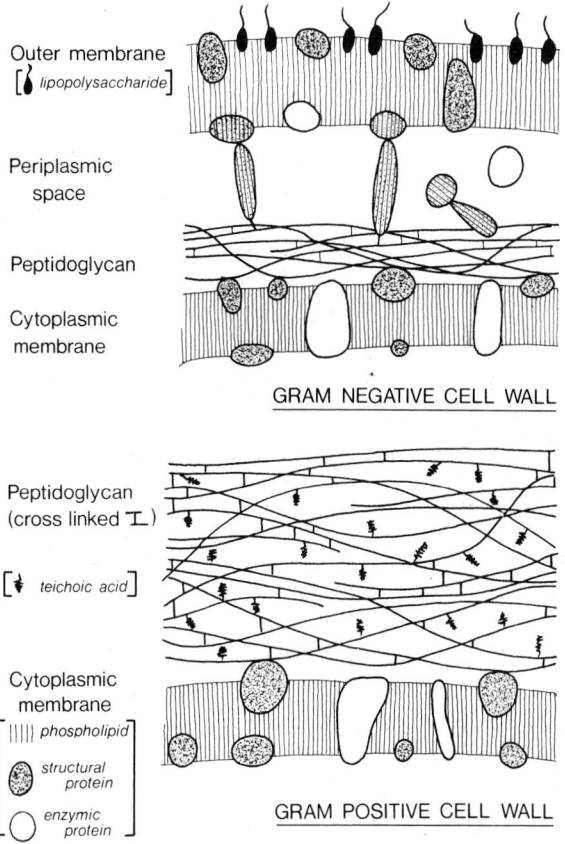

Outer membrane [● lipopolysaccharide]

Periplasmic space

Peptidoglycan

Cytoplasmic membrane

GRAM NEGATIVE CELL WALL

Peptidoglycan (cross linked ⊥)

[⸸ teichoic acid]

Cytoplasmic membrane [|||| phospholipid]

structural protein

enzymic protein

GRAM POSITIVE CELL WALL

Fig. 2.3 Diagram showing the structure of Gram-negative and Gram-positive bacterial cell walls.

2. *Spheroplasts:* derived from Gram-negative bacteria; retain some residual but non-functional cell wall material; osmotically fragile; produced by growth with penicillin and must be maintained in hyperosmotic medium.

3. *L-forms:* cell wall-deficient forms of bacteria usually produced in the laboratory but sometimes spontaneously formed in the body of patients treated with penicillin; more stable than protoplasts or spheroplasts, they can replicate on ordinary media.

4. *Mycoplasma:* an independent bacterial genus of naturally occurring bacteria which lack cell walls; also stable and do not require hyperosmotic conditions for maintenance.

a) LIPOPOLYSACCHARIDE

| Lipid A | ─KDO─Core polysaccharide─'O' Antigen sugar side chains |

Hydrophobic; embedded in outer membrane; endotoxin activity.

Hydrophilic; protrudes into environment.

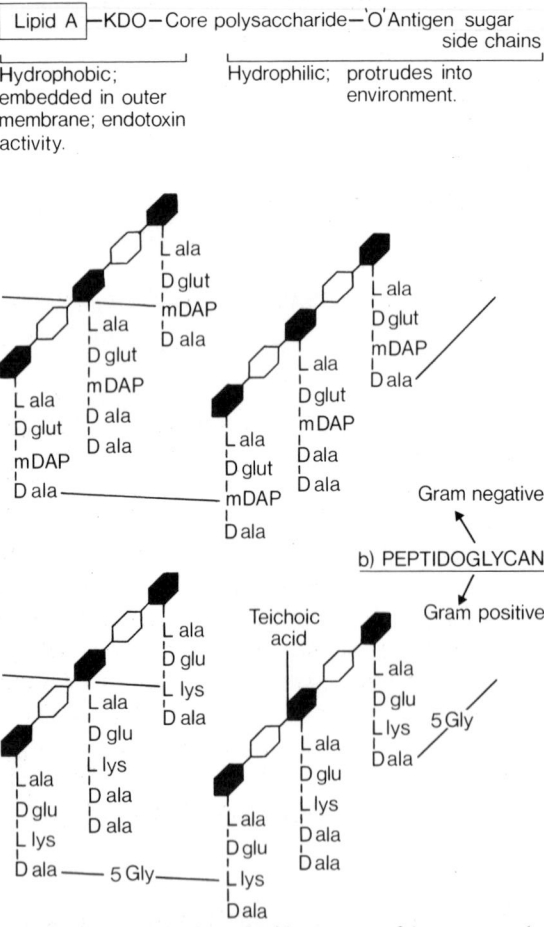

b) PEPTIDOGLYCAN

Fig. 2.4 Diagram of (a) the lipopolysaccharide structure of the outer membrane of the Gram-negative bacterial cell wall and (b) the detailed structure of the peptidoglycan component of the cell walls of Gram-negative and Gram-positive bacteria.

Key: ◆ = N-acetyl muramic acid; ◇ = N-acetyl glucosamine; L ala = L-alanine; D glu = D-glutamine; L lys = L-lysine; D ala = D-alanine; 5 Gly = pentaglycine; M DAP = meso-diamino-pimelic acid; D glut = D-glutamic acid.

Cytoplasmic membrane

A 'unit membrane' i.e. a double-layered structure composed of lipid and protein which acts as a semipermeable membrane through which there is uptake of nutrient by passive diffusion. It is also the site of numerous enzymes involved in the active transport of nutrients and various other cell metabolic processes.

Mesosomes
Convoluted invaginations of cytoplasmic membrane often at sites of septum formation: involved in DNA segregation during cell division and respiratory enzyme activity.

Nuclear material
The single circular chromosome which is the bacterial genome or DNA undergoes semiconservative replication bidirectionally from a fixed point — the *origin*.

Ribosomes
Ribosomes are distributed throughout the cytoplasm and are the sites of protein synthesis.

Inclusions
Sources of stored energy, e.g. polymetaphosphate (volutin), poly-β-hydroxybutyrate (lipid), polysaccharide (starch or glycogen).

Spores
Spores produced by bacteria in the genera *Bacillus* and *Clostridium* enable them to survive adverse environmental conditions: developed from and at the expense of the vegetative cell. Spores are dense, contain a high concentration of calcium dipicolinate and are resistant to heat, desiccation and disinfectants; they often remain associated with the cell wall of the bacillus from which they develop and are described as 'terminal', 'subterminal', etc. When growth conditions become favourable they germinate to produce vegetative cells.

TAXONOMY

Taxonomy is the classification or division of organisms into ordered groups.

Nomenclature is the labelling of the groups and of individual members within groups.

Organisms fall into three *kingdoms:*
1. Animals
2. Plants
3. Protista — contains all unicellular organisms including bacteria

Higher organisms are classified phylogenetically (i.e. on the basis of evolution) but this classification is impossible in the case of bacteria as they lack sufficient morphological features.

Bacterial classification is therefore artificial: different characteristics have been chosen arbitrarily so that the various members can be distinguished. These characteristics include:

Morphology
Staining
Cultural characteristics
Biochemical reactions
Antigenic structure
Base composition (i.e. GC ratio) of bacterial DNA

Tables 2.2 and 2.3 list most of the medically important bacterial genera classified on the basis of Gram's stain, morphology and aerobic or anaerobic growth: note that this classification has been simplified — for example, microaerophilic organisms are classified as anaerobes.

Table 2.2   Simplified classification of Gram-positive bacteria

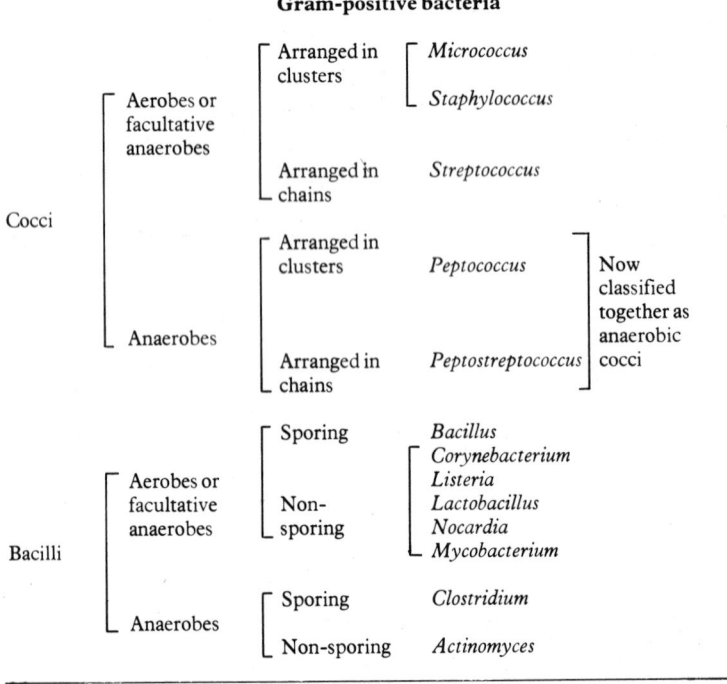

Table 2.3 Simplified classification of Gram-negative bacteria

		Gram-negative bacteria	
Cocci	Aerobes	*Neisseria*	
	Anaerobes	*Veillonella*	now classified with anaerobic cocci
Bacilli	Aerobes	*Pseudomonas*	
	Aerobes or facultative anaerobes	*Salmonella* *Shigella* *Klebsiella* *Proteus* *Escherichia* *Yersinia*	Enterobacteria
		Bordetella *Haemophilus* *Brucella* *Pasteurella* *Vibrio*	Parvobacteria
	Anaerobes	*Campylobacter* *Bacteroides* *Fusobacterium*	
Spirochaetes	Aerobes	*Leptospira*	
	Anaerobes	*Borrelia* *Treponema*	
Cell wall deficient bacteria		*Mycoplasma*	

3

Growth and nutrition of bacteria

Bacteria, like all cells, require nutrients for the maintenance of their metabolism and for cell division. Fast-growing bacteria divide approximately every 30 mins.

Chemically, bacteria consist of:

Protein
Polysaccharide
Lipid
Nucleic acid
Peptidoglycan

Bacterial growth requires:

1. materials for the synthesis of structural components and for cell metabolism
2. energy

Bacteria differ widely in their nutritional requirements. Some bacteria can synthesise all they require from the simplest elements. Others — including most pathogenic bacteria — are unable to do this: they need a ready-made supply of some of the organic compounds required for growth; other necessary compounds can be synthesised from breakdown products of complex macromolecules (e.g. proteins, nucleic acids) which are taken into the cell and degraded by bacterial enzymes. This is illustrated diagrammatically in Figure 3.1.

Elements. Bacterial structural components and the macromolecules used in cell metabolism are synthesised from the elements shown in Table 3.1; all these elements are therefore necessary for bacterial growth — whether available simply as elements or as part of complex molecules.

The four most important elements are:

1. Hydrogen
2. Oxygen
3. Carbon
4. Nitrogen

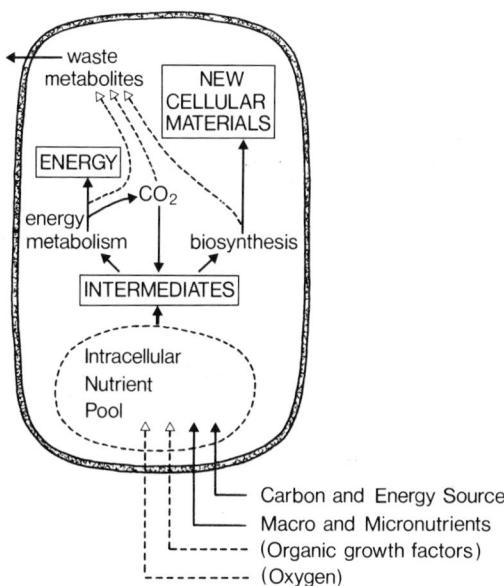

Fig. 3.1 Bacterial nutrition and metabolism.

Hydrogen and oxygen are obtained from water — essential for the growth and maintenance of any cell — and therefore available in any situation of potential bacterial growth.

Carbon and nitrogen are therefore the principal elements for which a source must be found.

Carbon

Depending on their requirements, bacteria can be classified as:

1. *Autotrophs:* free-living, non-parasitic bacteria, most of which can use *carbon dioxide* as their carbon source. The energy needed for their metabolism can be obtained from:
   a. Sunlight — photoautotrophs
   b. Inorganic compounds by oxidation — chemoautotrophs
2. *Heterotrophs:* generally parasitic bacteria; require more *complex, organic compounds* than carbon dioxide as their source of carbon and energy, e.g. sugars.

Sugars are degraded either by *oxidation* or *fermentation* (i.e. without oxygen): this provides energy for the generation of ATP, the universal energy storage compound.

Human pathogenic bacteria are heterotrophs: they are parasitic in that they have evolved to become adapted to an environment

Table 3.1 Bacterial growth — essential elements

Group	Element	Required for:
I Elements required for the synthesis of structural components	Carbon Hydrogen Oxygen Nitrogen Phosphorus Sulphur	Synthesis of: Carbohydrate and lipid, protein, nucleic acid
II Elements required for other cellular functions	Potassium Calcium Magnesium	Major cation; activates various enzymes Enzyme cofactor (e.g. proteinases); key role in spore formation Multi-enzyme cofactor, stabilises ribosomes, membranes, nucleic acid; enzyme-substrate binding
	Iron	Electron carrier in oxidation-reduction reactions; many other functions
III Trace elements	Copper Cobalt Manganese Molybdenum Zinc	Activators and stabilisers of a wide variety of enzymes

— i.e. the human body — in which many of their nutrients are both ready-made in complex form and freely available. Such bacteria have lost the biosynthetic mechanisms necessary for a free-living existence in a harsher environment, e.g. soil or water.

Nitrogen
The main source of nitrogen is ammonia, usually in the form of an ammonium salt: this may be available in the environment or produced by the bacterium as a result of deamination of amino acids.

Organic growth factors
These are organic compounds that cannot by synthesised by many bacteria; an exogenous supply is therefore required, although often only in small amounts.

Some examples are listed below:
1. *Amino acids* are required by many bacterial species which cannot synthesise them from simpler elements. Bacteria possess enzymes which degrade proteins to component amino acids; these then form an intracellular pool from which the appropriate amino acid is withdrawn to become incorporated into bacterial proteins.
2. *Purines and pyrimidines* are the precursors of nucleic acids and co-enzymes. In bacteria which require them, they are converted into nucleosides and nucleotides before incorporation into DNA and RNA.
3. *Vitamins.* Many pathogenic bacteria lack the ability to synthesise vitamins. Most vitamins are required for the formation of co-enzymes.

Prototrophs and auxotrophs
Prototrophs are *wild-type* bacteria with 'normal' growth requirements.
Auxotrophs are mutants which require an additional growth factor not required by the parental or wild-type strain: the growth factor may be many different types of chemical compound, e.g. an amino acid.

NUTRIENT UPTAKE

Most nutrients are small molecules which diffuse freely across the bacterial cytoplasmic membrane to enter the cell: some are at a higher concentration within the bacterial cell than in the external environment: their uptake is therefore an energy-dependent process.

Sugars are nutrients of relatively large size and therefore diffuse slowly.

Enzymes which facilitate the rapid uptake of larger nutrient molecules are present in many bacteria: usually associated with the cell membrane and energy-dependent, they may be *inducible* — i.e. produced only in the presence of the substrate or *constitutive* — i.e. produced constantly and independently of the substrate.

ENVIRONMENTAL CONDITIONS GOVERNING GROWTH

Water

Moisture is an absolute requirement for the growth of all bacteria: at least 80 per cent of the bacterial cell consists of water: the availability of water, for example, largely determines the size of the population of bacteria that can be supported by the skin.

Oxygen

Bacteria differ in their need of molecular oxygen for growth or in their need for its exclusion: this is illustrated in Table 3.2.

Table 3.2   Effect of oxygen on bacterial growth

Bacteria	Growth	
	In free oxygen	In absence of oxygen
Aerobes		
strict aerobes	+	−
facultative anaerobes	+	+
Anaerobes		
strict anaerobes	−	+
microaerophiles	−	+ (with trace of oxygen and carbon dioxide)

Carbon dioxide

Required by all bacteria and generally obtained as a product of their metabolism. Slow-growing or fastidious organisms may not generate enough carbon dioxide so that this must be supplied exogenously: this requirement becomes increased by *environmental stress,* e.g. caused by the transfer of bacteria from growth *in vivo* to culture *in vitro*.

Many pathogenic bacteria therefore require addition of 5 to 10 per cent carbon dioxide to the incubator atmosphere for *primary isolation* from clinical material.

Temperature
Bacteria also differ with regard to the optimal temperature range for their growth:

Psychrophile   below 20°C
Mesophile   between 25°C and 40°C
Thermophile   between 55°C and 80°C

Most medically important species are mesophiles and grow best at temperatures around 37°C (i.e. body temperature).

Hydrogen ion concentration
Not surprisingly, the optimal pH for bacteria that have evolved in association with man is similar to physiological pH — i.e. 7.2 to 7.4. A few species have evolved to become adapted to ecological niches where the pH is either higher or lower than normal.

BACTERIAL GROWTH AND DIVISION

Bacterial growth is the result of a balanced increase in the mass of cellular constituents and structures; the biosynthetic processes on which this increase depends are fuelled by energy.

Cell division is initiated when the increase in cellular constituents and structures reaches a critical mass.

Bacteria divide by *binary fission.*

Bacterial chromosome: a circular double-stranded DNA molecule. It replicates semiconservatively and bidirectionally: the replicated genome *segregates* between the daughter cells; cross walls then form as the parent cell divides into the two daughter cells.

Bacterial growth cycle
The growth cycle of a bacterial population (on transfer into fresh medium) is shown in Figure 3.2.

Four main phases can be recognised:

1. *Lag phase:* the bacteria do not divide immediately but undergo a period of adaptation with active macromolecular synthesis.
2. *Exponential (log) phase:* cell division then proceeds at a logarithmic rate determined by the medium and conditions of the culture.

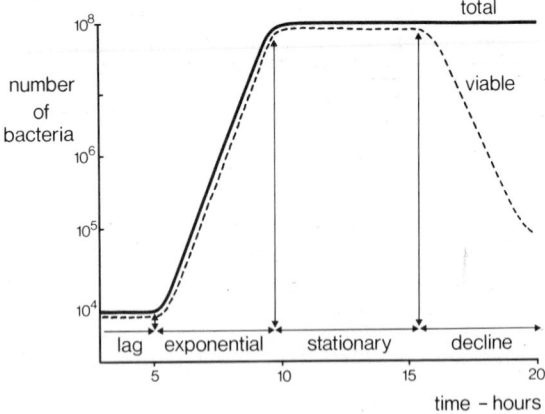

Fig. 3.2 The bacterial growth cycle.

3. *Stationary phase:* is reached when one or more essential nutrients become depleted; cell division ceases and there is no further growth.
4. *Decline phase:* after a period in the stationary phase, the bacteria start to die although the number of cells (viable and non-viable) remains constant.

This growth cycle is typical of a bacterial culture in conditions of *initial nutrient excess.*

During the exponential phase the population can double approximately every 30 min (with fast-growing bacteria): this is known as the *doubling time,* or *mean generation time,* and it can be calculated from the slope of the plot of the growth curve.

The plateau — i.e. the stationary phase — represents the *maximal cell density* or yield.

Figure 3.3 shows the effect of reducing the concentration of an essential nutrient. This results in a reduction in maximal cell density followed by a reduction in growth rate.

In open or continuous culture systems the bacterial population remains constant. A continuous supply of nutrient permits multiplication which is balanced by removal of bacteria. Under certain conditions one component of the nutrient medium can limit the population size and is known as the *growth limiting substrate.* The population is then *nutrient limited.*

Nutrient limitation governs the growth and maintenance of the normal bacterial flora of the human body.

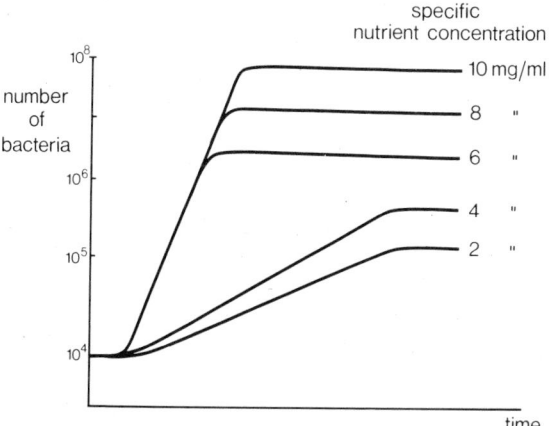

Fig. 3.3 The reduction in the yield of organisms and growth rate when the concentration of a specific nutrient in the medium is progressively decreased.

Growth in vivo

Bacterial growth in the human body is very different from that in artificial culture in the laboratory.

Bacterial growth *in vivo* resembles that of an open or continuous culture system in which the population has reached a *steady state* (i.e. both growth rate and population remain approximately constant).

A population like this is *nutrient limited* — the nutrient differing with the anatomical site. The nature of the limiting nutrient can influence bacterial activities such as invasiveness, toxin production or even antimicrobial susceptibility: it is therefore of considerable clinical importance.

Iron is an example of such a nutrient: iron limitation is common since iron is normally protein-bound to *transferrin:* bacteria therefore have to compete with transferrin to obtain sufficient iron for growth.

Iron limitation may play a role in preventing septicaemia i.e. when bacteria multiply in the blood.

4

Bacterial genetics

The study of variation in individual organisms, its origin and the subsequent behaviour of the variant bacteria in microbial populations.

ENVIRONMENT

Bacteria exist in a dynamic relationship with their environment. They constantly respond and adapt to it: the nature of the response, the adaptation and eventual evolution of the bacteria is a function of both organism and environment.

The environment determines whether a variant organism in a population survives: the environment subjects the population to survival stress — in other words, it applies *selective pressure.*

Selective pressure can be:

1. positive   — variants accumulate
2. negative   — variants are overgrown by the original population and disappear
3. neutral   — the original population and the variants survive and multiply together

Change in the environment can radically alter selective pressure (Fig. 4.1).

BACTERIAL VARIATION

There are two types of variation:
1. Physiological, non-heritable adaptation
2. Genetic, heritable variation

Physiological adaptation
Physiological adaptation to environmental changes produces *phenotypic variants:* this type of variation is reversible and involves no alteration in the cellular DNA. An example is shown

22

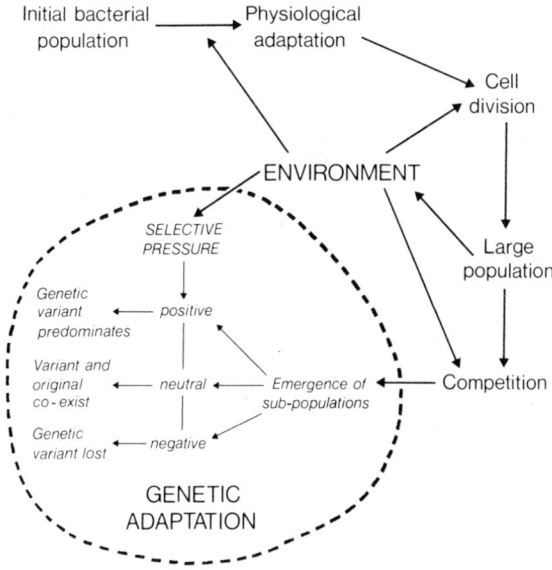

Fig. 4.1 The role of the environment in bacterial variation.

in Figure 4.2, which illustrates the classical inducer-repressor type of control in bacteria.

Genetic variation

Genetic adaptation involves alteration to the bacterial DNA.

Bacterial DNA

The genome or bacterial chromosome, is a circular molecule of double-stranded DNA: there is no nuclear membrane but the DNA is associated with the cytoplasmic membrane (Fig. 4.3a).

Replication: is semi-conservative and bidirectional starting from a fixed origin (Fig. 4.3b and c).

DNA synthesis: in bacteria involves complex and coordinated enzyme systems. *Repair mechanisms* also exist in most bacteria: these excise incorrect nucleotide sequences with nucleases, replace them with correct bases and religate.

Genetic variation can be brought about by:

1. mutation
2. gene transfer

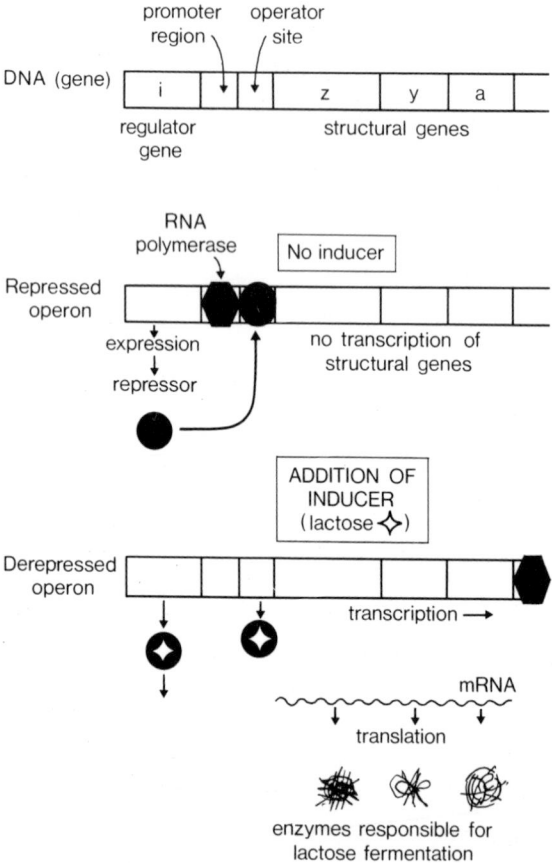

Fig. 4.2 Phenotypic adaptation. Lactose fermentation in *Escherichia coli* is an example of the regulation of enzyme synthesis at the level of transcription. In the absence of inducer (lactose) the 'i' gene product (the repressor) binds to the operator site and by acting as a barrier to movement of RNA polymerase from the promotor site prevents the transcription of the following structural genes. In the presence of inducer, lactose combines with the repressor which then dissociates from the operator site, to allow RNA polymerase to transcribe the structural genes of the operon with subsequent production of the enzymes involved in lactose fermentation.

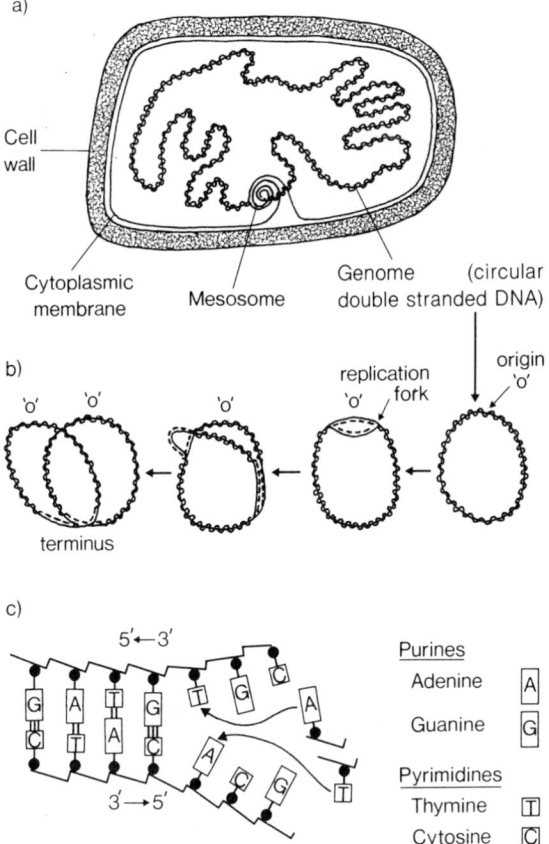

Fig. 4.3 Diagram showing (a) the bacterial genome, (b) bidirectional replication of bacterial DNA and (c) the molecular events at the replication fork.

1. MUTATION

Mutation is a chemical alteration in DNA. Mutants are variants in which one or more bases in their DNA are changed: this change is heritable and irreversible (unless there is back-mutation).

Mutants have a defect in a gene which results in altered transcription or an alteration in the amino acid composition of the protein which is the gene product.

Mutation can involve any of the numerous genes in bacterial DNA: many mutations are probably never detected since detection depends on the mutation affecting a recognisable function (e.g. causing antibiotic resistance).

Molecular basis of mutation

Mutation involves change in the sequence of bases in DNA: change of a single base alters the genetic code so that the triplet involved codes for a different amino acid: the new amino acid then becomes substituted for the correct amino acid in the protein product of the gene affected.

Mutation can be of three types:

1. *Base substitution:* change of a single base to one of the three other bases with consequent alteration in the triplet of the code. This may be:

   a. *Transition* in which purine/pyrimidine orientation is preserved, e.g. GC→AT

   b. *Transversion* with altered purine/pyrimidine orientation, e.g. GC→CG

2. *Deletion:* loss of a base affects the reading of subsequent triplets — frame-shift mutation (Fig. 4.4). Deletion sometimes involves several bases rather than a single base.

3. *Insertion* of an additional base also alters the reading frame of the DNA (Fig. 4.4).

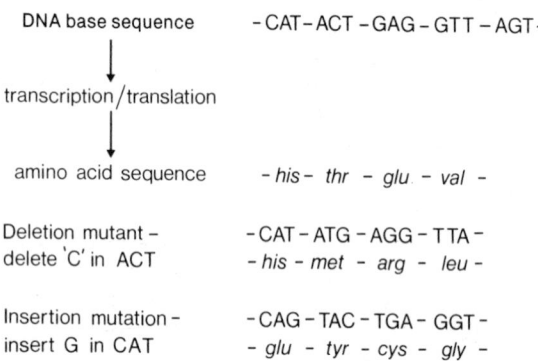

Fig. 4.4 Mutation. The effect of the deletion and insertion of a single base on the amino acid sequence of the gene product.

2. GENE TRANSFER

There are *three main types* of gene transfer that alter the DNA content of bacteria:

a. Transformation
b. Transduction
c. Conjugation

a. TRANSFORMATION

Fragments of exogenous bacterial DNA are taken up by recipient cells which they 'transform': transformation is detected by an alteration in the phenotype of the recipient bacteria and depends on *recombination* between one strand of the incoming double-stranded DNA fragment and the chromosome of the recipient bacterium. As in other bacterial systems, recombination depends on extensive DNA homology and on the function of the *rec A* gene. Transformation is illustrated diagramatically in Figure 4.5.

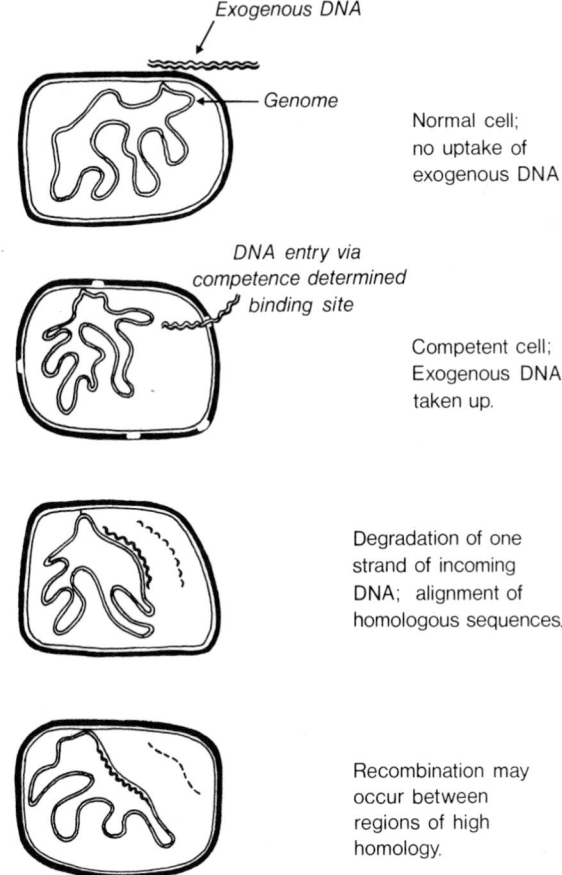

Exogenous DNA

Genome

Normal cell; no uptake of exogenous DNA

DNA entry via competence determined binding site

Competent cell; Exogenous DNA taken up.

Degradation of one strand of incoming DNA; alignment of homologous sequences.

Recombination may occur between regions of high homology.

Fig. 4.5   Transformation. Gene transfer by the uptake and subsequent recombination of a fragment of exogenous DNA.

Competence. The recipient bacteria must be competent — usually a transitory state in microbial cultures.

Frequency. The frequency of transformation is low.

Transformation is a laboratory manipulation and probably does not take place in nature.

b. TRANSDUCTION

The transfer of fragments of DNA from one bacterium to another by means of bacteriophage (or phage). The DNA from the donor bacterium is enclosed within a phage particle and is released on infection into a susceptible recipient bacterium (Fig. 4.6). The donated fragment may recombine with the recipient genome and the function transferred persists even if the phage

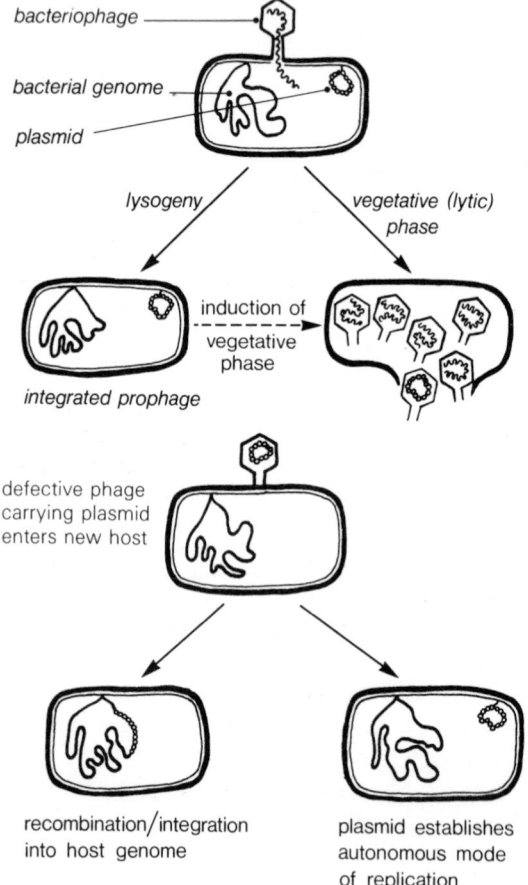

bacteriophage

bacterial genome

plasmid

lysogeny

vegetative (lytic) phase

integrated prophage

induction of vegetative phase

defective phage carrying plasmid enters new host

recombination/integration into host genome

plasmid establishes autonomous mode of replication

Fig. 4.6   Transduction. Gene transfer mediated by bacteriophage. Note that the bacterial genome of double-stranded DNA is shown as a single solid line.

genome is later lost. Plasmids (see below) are sometimes transferred in this way.

The mechanism of transduction is shown in Figure 4.6.

It is doubtful if this is a common mechanism of naturally occurring variation: it is probably responsible for the transfer of penicillin resistance which is plasmid-mediated in *Staphylococcus aureus*.

Phage conversion. Unlike transduction, with phage conversion the altered function of the recipient bacterium is lost when the phage genome is lost from the cell: in other words, the presence of the phage DNA is necessary for the maintenance of the function. Integretation of phage DNA into the bacterial genome acts as a switch to cause expression of otherwise unexpressed bacterial genes. Among the characters activated by phage conversion are toxin production in *Corynebacterium diphtheriae* and the formation of certain O antigens in salmonellae.

c. CONJUGATION

DNA is transferred from donor to recipient bacterium by direct contact probably via a tube or sex pilus (Fig. 4.7).

The DNA is usually in the form of an extrachromosomal element, or plasmid: sometimes the element is integrated into the donor chromosome and this is transferred.

Plasmids

Closed, circular molecules of double-stranded DNA which vary in size from 1.5 to 400 megadaltons: they replicate autonomously (i.e. they are replicons) usually in synchrony with the bacterial chromosome.

The genes coded by plasmid DNA are not essential for the survival of the host bacterium. Plasmids are sometimes lost from a cell: related plasmids cannot stably co-exist in the same cell.

Plasmids may code for:

(i) Their maintenance and replication

(ii) Transmissibility to other bacteria

(iii) Resistance to antibiotics

(iv) Production of bacteriocines (antibiotic substances which inhibit other strains of the same species).

F factor

The F or fertility factor or plasmid demonstrates one of the best studied examples of plasmid transfer by conjugation (Fig. 4.7).

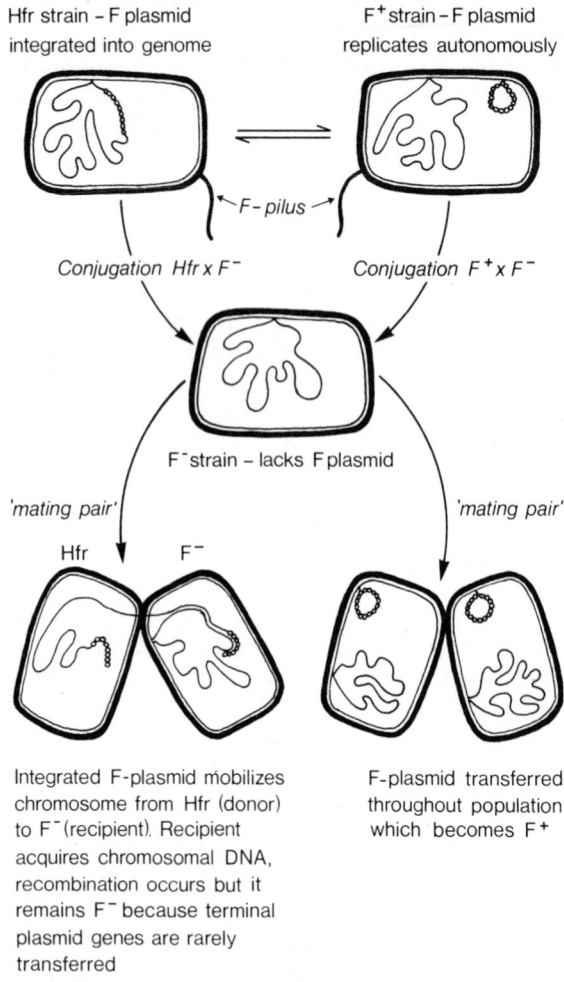

Fig. 4.7 Conjugation. Gene transfer by cell to cell contact.

The F factor codes for transmissibility and the production of a F or sex pilus: F$^+$ bacteria carrying the F factor can mate with F$^-$ bacteria which lack it: the F factor transfers to the F$^-$ recipient bacteria and these become F$^+$.

The F factor can integrate into the chromosome of the donor bacterium which then shows a high frequency of recombination (Hfr): when a Hfr bacterium conjugates with a F$^-$ cell, a single strand of chromosomal DNA — but not usually the F factor — is transferred into the recipient bacterium.

F' factors are those which have acquired by recombination some genes from the chromosome of the donor bacterium.

R or resistance factors are plasmids containing genes which code for antibiotic resistance: these may contain transfer genes (i.e. RTF or *resistance transfer factors*) which enable them to transfer, often readily, to other cells (Fig. 4.8). This method for the

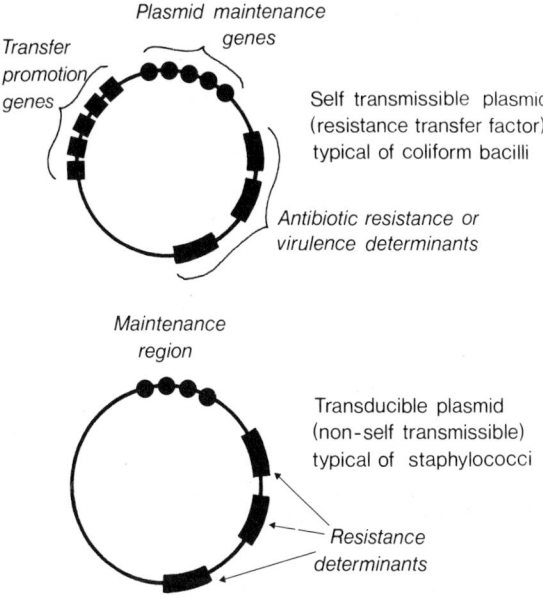

Fig. 4.8 The structure of two types of plasmid that code for antibiotic resistance in bacteria.

spread of transmissible antibiotic resistance is a serious problem in hospitals — see Chapter 46.

R factors which lack transfer genes are not transmissible unless present in a cell along with a plasmid that contains these genes.

TRANSPOSITION

Transposons

Transposons are segments of DNA that can move readily from plasmid to plasmid or from plasmid to chromosome (and vice versa). They consist of a unique sequence of DNA bounded by terminal inverted or direct repeated DNA sequences: the repeat sequences are responsible for the ability to insert and translocate.

Transposons can insert into the recipient DNA strand (certain sites are preferred) and extensive homology between the repeat sequences of the transposon (which are responsible for integration) and the site of insertion in the recipient DNA is not required: transposition differs from classical recombination in bacteria in being independent of the function of the *rec A* gene.

Transposition is a mechanism which permits a high rate of change in the genes contained in plasmids — with consequent rapid plasmid evolution.

Coding. Transposons code for toxin production, resistance to several antibiotics, e.g. ampicillin, trimethoprim and many other functions.

5

Laboratory methods

Laboratory diagnosis of bacterial infection depends on:
Direct demonstration of the organism in the specimen
Isolation of the infecting organism
Serology — the demonstration of antibody to the organism.
The most widely used method of diagnosis in medical bacteriology is isolation of the causal organism.
Laboratory methods in medical bacteriology can be divided into five categories:
1. Microscopy.
2. Culture.
3. Bacterial identification.
4. Tests of antimicrobial agents.
5. Serology.

1. MICROSCOPY
Preparations
1. *Stained films:* heat-fixed smears on slides of specimens or bacterial cultures stained by flooding with appropriate dyes.
   Observe: staining characteristics, shape, arrangement of bacteria, presence of cells e.g. polymorphs (pus cells).
2. *Unstained (wet) films:* drops of liquid specimens or fluid cultures placed on a slide and covered with a coverslip.
   Observe: for bacteria and cells: bacterial motility.
   Note: the refractive index of bacteria is similar to that of the fluids in which they are suspended and can only be seen by restricting the illumination of the standard light microscope.

Microscopes
1. *Standard light microscopy:* the main method used in diagnostic bacteriology.
   Magnification:
   a. *Stained films:* are examined with the oil immersion objective ($\times$ 100): using $\times$ 10 eyepieces the final magnification is $\times$ 1000.

b. *Wet films:* are examined with high power dry objective (× 40): final magnification with × 10 eyepieces is × 400.

Definition the ability to render the outline of the object clear and distinct depends on the quality of the lenses: good (but expensive) optical systems produce less aberration.

Resolution — the ability to distinguish two adjacent points as separate entities — is determined by the wave length of light: the best resolution obtainable is of the order of 0.25 μm.

2. *Dark-ground microscopy:* a special condenser illuminates the specimen obliquely so that light does not enter the objective: light is scattered by bacteria which appear brightly illuminated against a dark background.

3. *Phase contrast microscopy:* a special condenser and objective are used: direct light from the source and light scattered by structures in the field are transmitted so that when direct and diffracted beams unite they are not in phase: details of the object appear as differences in intensity and the contrast produced shows details of the fine structure of unstained, living microorganisms.

4. *Fluorescence microscopy:* uses ultraviolet light: bacteria or cells stained with auramine or other suitable fluorescent dyes alter the wave length and become visible as bright objects against a dark ground.

Immunofluorescence: combines serology with fluorescence microscopy by using antibody labelled with fluorescent dyes (e.g. fluorescein isothiocyanate, lissamine rhodamine) to detect specific antigens and so identify bacteria.

5. *Electron microscopy:* a beam of electrons allows resolution of extremely small objects e.g. 0.001 μm: the electrons are focused by electro-magnetic fields and the image is visualised — and can be photographed — on a fluorescent screen: used for study of the ultrastructure of bacteria.

Staining

1. *Gram's stain:* by far the most widely used stain in medical bacteriology: it not only reveals the shape and size of bacteria but enables them to be immediately classified into two categories — Gram-positive and Gram-negative.

Method
a. Crystal violet
b. Iodine solution
c. Decolourise with acetone or alcohol
d. Counterstain (e.g. dilute carbol fuchsin or neutral red)

Observe:

(i) *Gram-positive* bacteria resist decolourisation and stain blue-black.

(ii) *Gram-negative* bacteria stain pink with the counterstain since they are decolourised.

(iii) *Cells* are Gram-negative: although this staining method shows sufficient detail for polymorphs to be distinguished from other cells, it does not reveal much cytological detail.

2. *Ziehl-Neelsen's stain:* another important stain; used to demonstrate acid-fast bacilli.

Method
a. Concentrated carbol fuchsin mordanted by heating (with a swab soaked in alcohol and lit).
b. Decolourise — with 20 per cent sulphuric acid followed by 95 per cent ethanol.
c. Counterstain — methylene blue or malachite green.

Observe: for red bacilli against a blue background.

Mycobacteria — which include tubercle bacilli — are amongst the very few bacteria which resist this form of decolourisation (i.e. are acid-fast and alcohol-fast).

Medical bacteriologists use various special stains in addition to those listed above e.g. for demonstration of volutin granules (Albert's or Neisser's stain), capsules, spores, spirochaetes, flagella etc.

2. CULTURE

Bacteria — unlike viruses — can be grown *in vitro* on artificial, cell-free media: however they differ widely in their growth requirements so that many different kinds of media must be used in diagnostic bacteriology.

Growth requirements

The parasitic and pathogenic bacteria encountered in medical microbiology are heterotrophs: their nutritional and metabolic requirements are sophisticated: they need organic materials (carbohydrates, amino acids etc.) for growth. Many species in this category have to be provided with specialised growth factors for culture in the laboratory.

Media

Liquid media

Dispensed in tubes with cotton-wool stoppers or screw-capped bottles.

Growth is recognised: by turbidity in the fluid.

Uses:

1. Some bacteria, particularly if present in small numbers, will grow only in fluid media: occasionally a specimen contains inhibitory substances e.g. antibiotics, which are diluted by inoculation into a fluid medium thus allowing bacterial growth.
2. Fluid media with special constituents (e.g. sugars) are widely used to test the biochemical activities of bacteria for identification.
3. Enrichment media are fluids which encourage the preferential growth of a particular bacterium: they contain inhibitors for contaminants which might otherwise overgrow the pathogen.

Disadvantages:

1. Growth in fluid does not allow identification of bacteria by colonial morphology.
2. Fluid cultures mean that no estimate can be made of the numbers of bacteria originally present in the specimen.

Solid media

Dispensed in plastic or glass petri dishes, solid media are basically solidified fluid media.

Agar: the setting agent: added in a concentration of 1.5 per cent to give the medium the consistency of a firm jelly.

Method of inoculation: specimens or cultures of bacteria are plated or stroked out on the medium with a wire loop in such a way as to ensure a reducing inoculum: this means that after incubation, separated colonies will develop where individual bacteria have been deposited. (Fig 5.1a and b).

Uses:

1. The great value of a solid medium is that it allows separate colony formation resulting from the localised multiplication of the bacteria present in the specimen.
2. *Colonial morphology* enables most bacterial species to be presumptively identified.
3. *Quantitation:* the numbers and relative proportion of different bacterial species originally present in the specimen can be assessed.

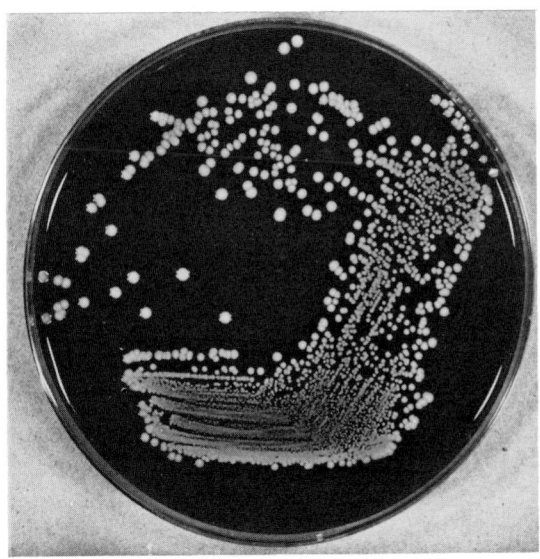

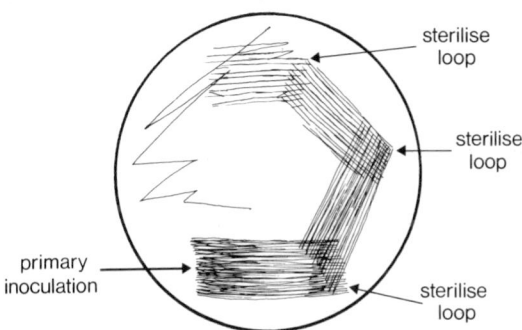

Fig. 5.1   Plating. Diagram illustrating the method of inoculating a plate of solid medium with bacteria to achieve separated colonies and a blood agar plate inoculated in this way after overnight incubation.

4. *Pure cultures* can be obtained by picking isolated bacterial colonies onto fresh solid medium: this is necessary for full identification.

   Selective media are solid media containing ingredients which inhibit unwanted contaminants (e.g. from the normal flora) but allow the desired pathogens to grow.

Common constituents of culture media

1. Water.
2. Sodium chloride, other electrolytes.
3. 'Peptone': a protein digest prepared from animal or vegetable protein by enzymic action; contains peptones, proteoses, amino acids, etc.
4. Meat extract, yeast extract: used to enrich media; contain protein degradation products, carbohydrates, inorganic salts, growth factors.
5. Blood: normally sterile, defibrinated horse blood: sometimes serum from various animals.
6. Agar: a carbohydrate derived from certain sea weeds: melts at 90°C but does not solidify until cooled to 40°C: heat sensitive ingredients can therefore be added just before the medium sets.

Nowadays media prepared from dehydrated ingredients are supplied by commercial firms: these are reconstituted and sterilised in the laboratory before use.

Table 5.1   Liquid media

Medium	Main constituents	Use
Peptone water	Peptone, sodium chloride, water	General culture; basal medium for sugar fermentation tests
Nutrient broth	Peptone water, meat extract	General culture
Glucose broth	Nutrient broth, glucose	Culture of delicate organisms
Robertson's meat medium	Nutrient broth, minced meat	Culture of anaerobic and aerobic bacteria
Tetrathionate broth	Nutrient broth, sodium thiosulphate, iodine	Enrichment culture for salmonellae
Selenite F broth	Peptone water, sodium selenite	Enrichment culture for salmonellae and shigellae

Table 5.2 Solid media.

Medium	Main constituents	Use
Nutrient agar	Nutrient broth, agar	General culture
Blood agar	Nutrient agar 5 to 10% horse blood	General culture: the most widely used medium in medical bacteriology
Chocolate agar	Heated blood agar	Richer than blood agar: isolation of H. influenzae, N. gonorrhoeae
MacConkey agar	Peptone water agar, bile salt, lactose, neutral red	Culture of enterobacteria: lactose-fermenting colonies are coloured pink
CLED agar (cystine-lactose electrolyte deficient medium)	Peptone, L-cystine, lactose, bromthymol blue	Culture of enterobacteria: lactose-fermenting colonies are coloured yellow
Desoxycholate citrate agar	Nutrient agar, sodium desoxycholate, sodium citrate, lactose, neutral red	Selective medium for salmonellae, shigellae
Lowenstein-Jensen*	A mineral salt solution, glycerol, malachite green, whole egg	Culture of Myco. tuberculosis
Antibiotic sensitivity agar	Peptone in a semi-synthetic medium designed to avoid antibiotic inhibitors	Antibiotic sensitivity tests

* *Note.* This medium is rendered solid by heating which 'sets' the egg: it does not contain agar

Hydrogen-ion concentration: the final pH of most media used in diagnostic bacteriology is between 7.0 and 7.6.

Tables 5.1 and 5.2 list some of the most widely used liquid and solid media.

Media for blood cultures

Two bottles are supplied:

1. Nutrient broth for aerobic culture: sometimes with an agar slope set onto the narrow side of the bottle so that colonies can be observed.
2. Nutrient broth with sodium thioglycollate (a reducing agent) or Robertson's meat medium: for anaerobic culture.

The bottles have rubber seals and a perforated metal cap: inoculated by injection of aseptically-collected blood through the hole in the cap with a syringe.

Transport medium

Aids preservation of delicate pathogens during transit to the laboratory.

Stuart's Transport Medium: a semi-solid, non-nutrient agar with thioglycollic acid (as reducing agent) and electrolytes. Originally devised for *Neisseria gonorrhoeae* — now used as a general transport medium.

Incubation

1. *Atmosphere*
   a. Most human pathogens grow in air: the addition of 10 per cent carbon dioxide is essential for the isolation of some species and enhances the growth of many others.
   b. *Anaerobic bacteria* require incubation in an atmosphere without oxygen: petri dishes are placed in a sealed jar from which oxygen is removed by combination with hydrogen to form water in the presence of a catalyst (e.g. alumina pellets coated with palladium): hydrogen and carbon dioxide are liberated inside the jar by means of a commercially available gas generating system.

2. *Temperature*
   The optimal temperature for the growth of most pathogens is body heat — 37°C: however a few bacteria require a higher and some a lower temperature.

3. BACTERIAL IDENTIFICATION

Bacteria isolated from a specimen on culture must next be identified.
   This proceeds in the following stages:

1. Colonial and microscopic morphology

The appearance of bacterial colonies is usually very characteristic and allows presumptive identification to be made on primary cultures: a stained film of the colony may be required for identification: however most species require confirmatory tests.

2. The conditions required for growth
Also aid identification, e.g. aerobic or anaerobic growth, ability to grow on simple or only on enriched media etc.

3. Biochemical tests
The main tests used are listed in Table 5.3: these are principally used for enterobacteria.

Table 5.3   Biochemical tests used in bacterial identification

Test	Substrate	Observe
Carbohydrate metabolism		
Breakdown of sugars a. Oxidative (aerobic) b. Fermentative (anaerobic)	Various carbohydrates	Acid, sometimes with gas
Voges–Proskauer	Glucose	Acetyl methyl carbinol
Citrate utilisation	Sodium citrate	Growth (citrate is the sole carbon source)
Protein metabolism		
Gelatin liquefaction	Gelatin	Liquefaction
Amino acid decarboxylases	Lysine, ornithine, arginine	Carbon dioxide
Production of hydrogen sulphide	Sulphur-containing amino acids	Hydrogen sulphide
Phenylalanine deaminase	Phenylalanine	Phenylpyruvic acid
Indole	Tryptophan	Indole
Other tests		
Urease	Urea	Ammonia
Catalase	Hydrogen peroxide	Oxygen
Oxidase	Redox dye	Oxidation to purple colour
Nitrate reduction	Potassium nitrate	Nitrite

Note. Most of these tests are so devised that the products result in the development of a visible change e.g. indicators for pH change, gas production etc.

API system: nowadays most laboratories use commercially prepared kits which enable a range of tests to be carried out rapidly and accurately: the most widely used is the API 20E for the identification of aerobic Gram-negative bacilli (Fig. 5.2).

Fig. 5.2   API strip. A commercial kit of 20 biochemical tests used to identify enterobacteria.

Method: after inoculation and incubation the results are scored to give a numerical profile: this is compared statistically to a profile compiled from type cultures and the degree of correspondence enables identification to be made with a known degree of certainty.

4. Recognition of enzymes
The production of enzymes is the basis of most of the reactions included in the biochemical tests in Table 5.3. However, some bacteria other than enterobacteria can be identified primarily by production of a characteristic enzyme.

   a. *Coagulase:* clots plasma (specific for *Staphylococcus aureus*)
   b. *Lecithinase:* produces opacity in egg or serum medium (if inhibited by *Clostridium welchii* antitoxin, identifies this organism — the Nagler Reaction)

5. Antigenic structure
The definitive method of identification: depends on recognition of a variety of protein and polysaccharide antigens in flagella, cell wall, capsule or liberated from the bacteria as toxins. Particularly useful for the large numbers of biochemically-similar enterobacteria.
Methods: various immunological techniques are used e.g. agglutination, precipitation, sometimes neutralisation of toxin.

6. Bacterial typing
It is often necessary for epidemiological reasons to identify individual strains — or types — of bacteria *within a species.* This can be done in the following ways:

   a. *Antigenic typing:* in which the antigens which distinguish different strains of bacteria are identified.

b. *Bacteriophage typing:* different strains of an organism have different susceptibilities to a series of bacterial viruses or bacteriophages: usually patterns of lysis with more than one phage are detected and can be used to distinguish one bacterial strain from another.

c. *Bacteriocine typing:* bacteriocines are proteins released by bacteria which inhibit the growth of other members of the same species: bacterial strains can be differentiated on the basis of the bacteriocins they produce by observing growth inhibition of a series of test strains.

4. TESTS OF ANTIMICROBIAL AGENTS

TESTS OF SENSITIVITY TO ANTIMICROBIAL AGENTS

One of the most important functions of a diagnostic bacteriology laboratory and a large part of the day-to-day workload.

There are two principal methods:

1. **Disc diffusion**

   By far the most widely used method.

   Method: paper discs impregnated with antibiotic solutions (at a concentration related to blood or urine levels) are placed on the surface of a plate inoculated all over with either the specimen or the bacterial culture under test.

   Observe: for zones of inhibition round the discs after incubation (Fig. 5.3).

   Primary sensitivity testing: in which the specimen is inoculated directly onto a sensitivity plate is a routine in some laboratories. Often successful with infected urines it may fail to give a readable result with other specimens.

   Secondary sensitivity testing: in which the inoculum is the bacterium isolated from the specimen is preferred by other laboratories: it is essential when the growth is mixed. Unfortunately, it involves a day's delay in reporting.

2. **Tube dilution**

   Laborious: only done in special circumstances e.g. in tests on bacteria from cases of infective endocarditis where treatment is difficult.

   Method: tubes with doubling dilutions of the antibiotic in broth are inoculated with the bacterium under test.

   Observe: for tubes in which bacterial growth is inhibited after

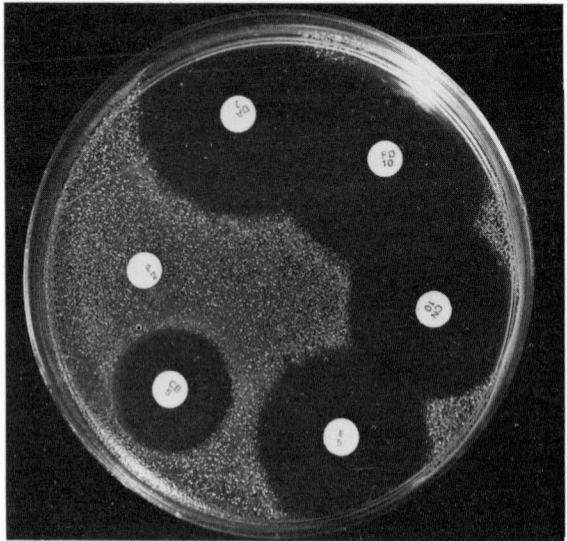

Fig. 5.3 Antibiotic sensitivity test showing zones of inhibition of the growth of the test organism round discs containing antibiotics to which it is sensitive. Resistance is shown by the growth of the organism right up to the disc.

incubation: the lowest concentration in which there is no growth represents the MIC or minimal (bacteriostatic) inhibitory concentration.

MBC or minimal bactericidal concentration is estimated by subculture from the tubes of a MIC test onto solid media: growth on subculture indicates the presence of surviving bacteria and that the concentration has not been bactericidal: estimated as the lowest concentration of antibiotic in which the bacteria have been killed (i.e. no growth on subculture from the appropriate tube).

Tests of combined antibacterial action
Carried out by the tube dilution method but using combinations of two antibiotics in concentrations corresponding to blood levels: usually only done when indicated clinically e.g. in infective endocarditis due to a resistant organism.

ESTIMATION OF LEVEL OF ANTIMICROBIAL AGENTS IN BLOOD

Now being increasingly used for two purposes:
1. To check that the concentration of drug in the blood is below that associated with toxicity

2. to confirm that the blood contains an adequate therapeutic level of the drug
   Method: specimens of serum are taken before and after a dose to estimate 'trough' and 'peak' levels respectively. They can be tested in one of two ways:
   a. *Bioassay:* by measuring zones of inhibition produced by the serum in wells in an agar plate inoculated with a suitable bacterial culture: these are compared to zones produced by standard concentrations of the drug and the level in the serum calculated.
   b. *Chemical or immunochemical assays:* the number of antimicrobial drugs that can be assayed by methods which give rapid and accurate results is increasing: commercially prepared kits are now available for some.

5. SEROLOGY

Some diseases — fewer now than formerly — are diagnosed by demonstration of antibody to the causal organism in the patient's blood.

Titre: is the term for the highest dilution of serum at which antibody activity is demonstrable: usually expressed as the reciprocal of the serum dilution e.g. 64 if antibody was detected at a final serum dilution of 1 in 64.

Methods
The following are the main immunological techniques used:

1. **Agglutination**
   Antibody in patient's serum is detected by its ability to cause visible aggregation of suspensions of bacteria.
   Indirect (Coombs') Agglutination: sometimes the antibody in a patient's serum is incomplete and although it combines with bacteria does not cause them to agglutinate: addition of rabbit anti-human globulin will then cause the antibody-coated bacteria to agglutinate.

2. **Precipitation**
   The antigen is in soluble form and antibody is detected usually by formation of a visible line of precipitate as a result of diffusion in agar gels.

3. **Complement fixation**

Combination of antigen with antibody 'fixes' or uses up complement: when an indicator system (sheep erythrocytes coated with rabbit antibody) is then added haemolysis takes place if there is free complement.

Observe:

Absence of haemolysis: indicating complement fixation due to antibody in the patient's serum having reacted with the original antigen — a positive result.

Presence of haemolysis: due to persistence of complement — a negative result.

4. **Immunofluorescence**

Usually indirect immunofluorescence in which smears containing the organism (antigen) are exposed to patient's serum followed by anti-human globulin labelled with a fluorescent dye. Serum antibody that has reacted with antigen attaches the labelled anti-human globulin: when viewed by ultra-violet microscopy the antigen fluoresces.

5. Recently developed techniques in immunology are beginning to be introduced into diagnostic bacteriology: these include
   a. **Radioimmunoassay (RIA)**
   b. **Enzyme-linked immunosorbent assay (ELISA)**

6

Sterilisation and disinfection

Doctors must know how to render articles safe from the risk of transmitting infection. This does not always require *sterility* which means the complete destruction of all organisms including spores. The degree of bacterial killing which is necessary depends on the circumstances. For example, pasteurised milk is far from sterile but is safe because any human pathogens of bovine origin that it might contain have been killed.

STERILISATION

Involves rendering an article sterile.

Bacteria in the vegetative — or non-sporing — state are readily killed by heat, e.g. at 56°C for 30 min or at 100°C for a few seconds.

Spores are survival mechanisms possessed by certain bacteria (in the field of medical microbiology all are members of the genera *Bacillus* and *Clostridium*) and — not surprisingly — are very resistant to heat and other inactivating agents: they are destroyed only by longer and more intensive application of the various methods used to kill bacteria.

Sterilisation is therefore complex, difficult and often costly: it is dependent on the following factors:

1. Knowledge of the killing or death curves of representative bacteria and spores when exposed to the inactivation process. Spores vary in the degree of their resistance to heat and other agents: each process should be tested against spores of appropriate resistance (*Bacillus stearothermophilus* for heat, *B. subtilis var. globigii* for ethylene oxide, *B. pumilus* for ionising radiation).
2. The penetrating ability of the inactivating agent — important in the case of surgical dressings or pads: steam penetrates much more effectively than dry heat.
3. The ability of the article to withstand the sterilisation process. Surgical fabrics tolerate steam but are destroyed by dry heat; plastics are heat-sensitive.

4. The effect of organic matter such as blood, faeces, dust, etc. which enhances the survival of bacteria and spores and interferes with the sterilisation process. Articles must be clean before they are sterilised.

Safety margin: sterilisation (and disinfection) must, in practice, always have a safety margin over and above the minimal treatment required for killing of bacteria and spores: this is arbitrarily fixed at 25 per cent and is to allow for possible inadequacies in the sterilisation process or for the presence of a few exceptionally resistant organisms or spores.

The various methods of sterilisation are shown in Table 6.1.

Table 6.1 Methods of sterilisation

Method	Equipment	Use
Moist heat (steam under pressure), e.g. at 121°C for 15 min or 134°C for 3 min	Autoclave	Surgical dressings, instruments, almost any article or fluid which is not heat-sensitive
Dry heat 160°C for 1 h	Hot air oven	Glassware, powders, ointments
Ethylene oxide (gaseous)	Special chamber	Plastic and rubber goods; certain sophisticated instruments; poor penetration therefore articles must be clean
Gamma irradiation	Carefully shielded cobalt-60 source. Expensive, special plant required. Used commercially	Plastic goods
Subatmospheric steam at 80°C with formaldehyde vapour*	Special chamber	Heat-sensitive equipment, e.g. endoscopes
Filtration through cellulose membranes	Filtration apparatus	Heat-sensitive fluids (will not remove viruses)

*Subatmospheric steam without formaldehyde is often used to render equipment safe by killing vegetative bacteria.

Autoclaves
Steam is a very efficient sterilising agent because when it condenses to form water it:

1. Liberates latent heat which participates in bacterial killing.
2. Contracts in volume enhancing penetration.

Protein coagulates (and bacteria are killed) more readily in the presence of moisture.

When water is heated within a closed vessel the temperature at which it boils, and that of the steam it forms, rises above 100°C. Thus *steam temperature* rises with increase in atmospheric pressure: e.g. at 15 lb per in² the temperature of steam is 121°C, at 30 lb, 134°C: conversely at subatmospheric pressure the temperature of steam is below 100°C. The time of exposure necessary for sterilisation diminishes with rising temperature, e.g. at 134°C only 3 min is necessary.

The simplest autoclave is a domestic pressure cooker. It suffers two defects: it is impossible (i) to dry the load, (ii) to expel all the air because air is denser than steam.

An *autoclave* consists of a double-walled or jacketed chamber: steam circulates within the jacket and is supplied under pressure to the closed inner chamber in which the goods for sterilisation are placed (Fig. 6.1).

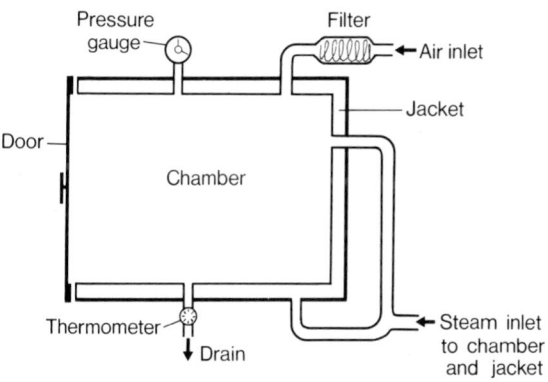

Fig. 6.1 Diagram of an autoclave (simplified).

Drying is accomplished at the end of the cycle by evacuating the steam from the chamber which contains the load while steam continues to circulate in the jacket so that the chamber remains hot. Moisture in the load evaporates and drying is assisted by sucking warmed filtered air into and through the chamber.

Expulsion of air is necessary because

1. The temperature of an air–steam mixture at a given pressure is lower than that of pure steam (Table 6.2).
2. Air pockets interfere with steam penetration.

Table 6.2 Autoclave pressures and temperatures

Chamber pressure	Chamber temperature of:		Usual method of air removal	Type of autoclave
	Pure steam	50 : 50 air– steam mixture		
15 lb	121°C	112°C	Gravity displacement by steam	Downward displacement
30 lb	134°C	128°C	Use of efficient pump	High vacuum

Downward-displacement autoclaves: obsolescent and relatively inefficient; a long cycle time is unavoidable. Although a preliminary vacuum may remove around half of the air, the remainder is displaced by admitting steam at the top of the chamber and allowing the heavier air to gravitate to the bottom where it is discharged. Only when this is complete can the sterilisation holding period commence: its duration depends on the time taken to penetrate the load and this is dictated by its composition and method of packing.

High-vacuum autoclaves: modern, fast and reliable equipment. Before the admission of steam, 98 per cent of the air present in the chamber is rapidly removed by an electric pump and sterilisation can proceed without delay. The cycle is short: all loads are penetrated efficiently.

Centralised sterilisation facilities using high-vacuum autoclaves are available to most hospitals today.

1. *Central Sterile Supply Department (CSSD):* from which packs of sterilised dressings, equipment, etc. are prepared and transported to the various wards, clinics and theatres: most — sometimes all — the sterilisation in the hospital is carried out in autoclaves in the CSSD by specially trained and supervised staff.
2. *Theatre Sterile Supply Unit (TSSU):* usually a subdepartment of CSSD specialising in theatre equipment, e.g. pre-set trays of surgical instruments.

Other sterilisation methods
The following are used:

Dry heat: by microbiology laboratories for glassware.

Ethylene oxide gas, gamma irradiation: largely by commercial suppliers for plastic goods.

Filtration: by pharmaceutical firms for the sterilisation of drugs for injection.

STERILE DISPOSABLE PLASTIC ARTICLES

Syringes, catheters, tubing, infusion bags etc. are now supplied as pre-sterilised disposable articles. The ready availability and comparative cheapness of these has revolutionised medical practice both in hospitals and general practice: they have also greatly improved safety since amateur attempts at sterilisation (especially of syringes) are no longer necessary.

Boilers

The small electrically heated boiler is a familiar piece of equipment in most doctors' and dentists' surgeries. The maximum temperature is, of course, 100°C and will therefore not kill all spores. Nevertheless it is perfectly satisfactory for rendering safe from the risk of infection metal instruments such as specula, tongue depressors, forceps, etc. It is not suitable for syringes.

DISINFECTANTS AND ANTISEPTICS

Disinfectants and antiseptics are chemicals which inactivate vegetative bacteria but are rarely capable of killing spores. Most bacteriologists regard them as unreliable since their efficacy depends on exposure for lengthy periods of time and can be reduced or abolished by the presence of blood, faeces or other organic matter. Nevertheless they are extremely widely used in medicine.

Disinfectants: crude corrosive chemicals that are destructive to skin and mucous membranes; generally more active antimicrobial agents than *antiseptics* which can be applied to body surfaces.

Table 6.3 lists the most common agents: proprietary names are shown in brackets but this list includes only the preparations most familiar to the authors; many other equally effective proprietary-name brands are available.

Bedpans

Safe disposal of faeces is a major problem in hospitals where a considerable proportion of the patients are confined to bed. Two main methods are now used:

1. Disposable papier-mâché bedpans: placed in a special apparatus which destroys the pan and flushes away the contents.

Table 6.3 Disinfectants and antiseptics

Disinfectant	Antibacterial spectrum	Inactivation by organic matter	Comments
Phenolics			
1. Clear fluids (Stericol, Sudol, Hycolin)	Wide	−	Irritant to skin
2. Chloroxylenol (Dettol)	Gram-positive bacteria	+	Not very effective, wide domestic use
3. Hexochlorophane (Gamophen soap, pHiso-Med)	Gram-positive bacteria	+	Often incorporated in soap or detergent solution, effect is cumulative
Hypochlorites (Chloros, Diversol Bx, Milton, Eusol)	Wide, including viruses	+	Corrosive to metal, kills some spores
Povidone-iodine (Betadine)	Wide	−	Hypersensitivity may be a problem
Formaldehyde	Wide, including viruses	−	Some effect on spores; used as gas or solution; irritant to eyes, respiratory tract
Glutaraldehyde (Cidex, Asept)	Wide, including viruses	−	Kills spores — slowly; penetration poor
Quaternary ammonium compounds, e.g. cetrimide (Cetavlon)	Gram-positive bacteria	+	Bacteriostatic, inactivated by soap and man-made materials
70% alcohol (ethyl or isopropyl)	Wide	−	Often used in combination with iodine or chlorhexidine
Chlorhexidine (Hibitane)	Gram-positive bacteria	+	Often used in combination with cetrimide (Savlon)

2. Stainless steel bedpans: locked into a disposal unit which empties them and flushes them out with steam.

Table 6.4 Hospital disinfection procedures

Article	Disinfectant
Floors, walls*	Clear phenolic fluids — as 1% solution — 2% if contaminated with pus or faeces
Surfaces, e.g. trays, locker tops, sinks	Hypochlorite or 70% alcohol
Skin 1. Surgeons' hands	Hexachlorophane soap, chlorhexidine in alcohol, povidone-iodine
2. Patient's skin at operation	Chlorhexidine in alcohol, iodine in alcohol, povidone-iodine
3. Patient's skin before venepuncture	70% alcohol
4. Cleansing of wounds, burns, pressure sores, ulcers	Aqueous solutions of chlorhexidine and cetrimide either alone or in combination; hypochlorites (Milton, Eusol)
Anaesthetic equipment, endoscopes	Gluteraldehyde Subatmospheric steam
Thermometers	Wipe with 70% alcohol; keep dry for individual patient

* A high standard of general domestic cleanliness is essential: the use of disinfectants is of secondary importance.

PUBLIC HEALTH ASPECTS — SAFE MILK AND WATER SUPPLIES

Pasteurisation of milk
One of the most successful applications of bacteriology: milk is raised to a temperature of either 63–66°C for 30 min or — in the flash method — to 72°C for 15 sec. Milk so treated is not sterile, as anyone who keeps it for a few days at room temperature will discover: but it is safe from contamination with viable *Mycobacterium tuberculosis*, brucellae, campylobacter, *Coxiella burneti* and other pathogenic vegetative bacteria.

Pasteurised milk is not spoiled in taste or appearance but the cream layer is slightly reduced. Almost all milk in Britain is pasteurised — milk-borne cases of infection (e.g. the outbreak of 98 cases of *Salmonella typhimurium* food-poisoning near Edinburgh in the autumn of 1978) are generally due to rural unpasteurised milk supplies.

Treatment of water

The natural habitat of some saprophytic bacteria is water and many soil bacteria gain access to water during heavy rain.

In addition, microorganisms of faecal origin from the human and animal intestine may find their way into water supplies: it is these that cause outbreaks of water-borne infection and the diseases spread in this way include enteric fever, dysentery, cholera, hepatitis A and gastroenteritis (sometimes without a recognised causal microorganism).

The number of bacteria in clean water in a good catchment area decreases rapidly on *storage*. Subsequently, before entering the piped supply, reservoir water is treated by

1. *Filtration* through sand supported on gravel and clinker
2. *Chlorination*

Medically important bacteria

7

Staphylococcus and micrococcus

Gram-positive cocci which are arranged in grape-like clusters.
Staphylococcus: pathogenic or commensal parasites.
Micrococcus: free-living saprophytes.

STAPHYLOCOCCUS

Species:
Staphylococcus aureus
Staphylococcus epidermidis (Staphylococcus albus)
Staphylococcus saprophyticus (previously classified as micrococcus subgroup 3).
The two main species are *Staph. aureus* — the major pathogen responsible for pyogenic infections — and *Staph. epidermidis* — a universal skin commensal.
Staph. saprophyticus: very similar in general properties to *Staph. epidermidis* — it can be identified by its *resistance to novobiocin.*
Habitat: the body surfaces and, by dissemination, air and dust.
1. *Staph. aureus:* carried in the anterior nares (50-75 per cent of healthy people); less often in the skin (especially axilla and perineum) and mucous membranes (throat, gut). Originating from these reservoirs *Staph. aureus* is ubiquitous and always present in the hospital environment.
2. *Staph. epidermidis:* normally present in the resident skin flora; also found in gut, upper respiratory tract.

LABORATORY CHARACTERISTICS

Morphology and staining. Gram-positive, non-sporing, non-motile cocci (diamater about 1 μm) characteristically arranged in clusters.
Culture: grow well on ordinary media aerobically and, although less well, anaerobically; optimal temperature 37°C.
Colonial appearance:
Staph. aureus: typically golden colonies but pigmentation varies from orange to white.

Staph. epidermidis: white colonies.

Pigment production with *Staph. aureus* is often poor after 24 h incubation at 37°C on nutrient agar or blood agar; this important differential feature can be enhanced by leaving the plates in the light at room temperature. Pigmentation is encouraged by growth on special media, e.g. milk agar.

Selective media: staphylococci will tolerate sodium chloride in concentrations of 5 to 10 per cent. Salt-containing agar and broth are useful in isolating staphylococci from samples likely to contain large numbers of other bacteria.

Identification of Staph. aureus: since pigment production is an uncertain character, a number of tests have been developed to distinguish *Staph. aureus* from *Staph. epidermidis.*

1. *Coagulase test:* the conclusive test that distinguishes *Staph. aureus* (coagulase positive) from *Staph. epidermidis* (coagulase negative).

   a. *Tube test.* Inoculate diluted human or rabbit plasma with the staphylococcus and incubate for 3–6 h at 37°C.
   Observe: clotting (gel formation) with coagulase-positive strains.
   Mechanism: extracellular *free coagulase* produced on culture has thrombin-like activity and converts fibrinogen into fibrin.

   b. *Slide test.* Emulsify the staphylococcus in a drop of saline and mix with a loopful of undiluted plasma.
   Observe: rapid clumping of the suspension with coagulase-positive strains (Fig. 7.1).

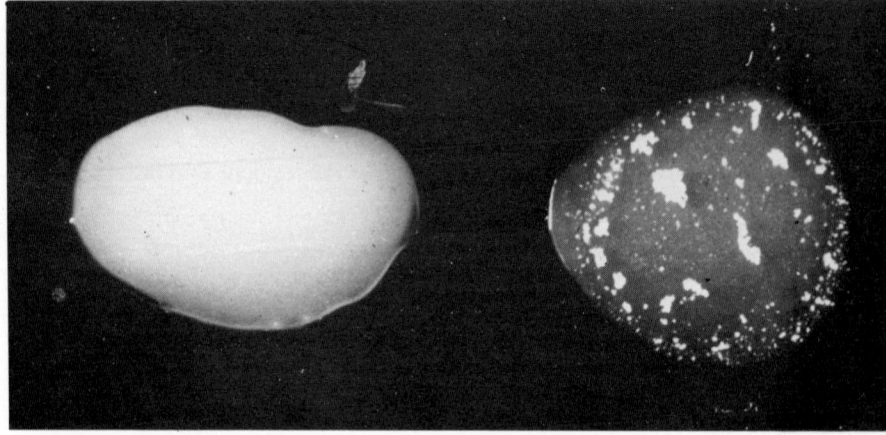

Fig. 7.1   Slide coagulase test. Bound coagulase in the strain of *Staphylococcus aureus* being tested has caused clumping with citrated plasma in the drop on the right; the left is the control drop of a uniform suspension in saline.

Mechanism: bound coagulase (clumping factor) on the bacterial cell surface reacts directly with fibrinogen.

Although the reactions are due to two entirely different substances the vast majority of strains give either a positive or a negative result with both tests.

2. *Phosphatase test. Staph. aureus* produces a phosphatase; very few coagulase-negative strains are phosphatase-positive.
3. *Deoxyribonuclease test.* Coagulase-positive strains hydrolyse DNA; only a minority of coagulase-negative strains produce a DNAase.
4. *Mannitol fermentation.* Almost all *Staph. aureus* strains form acid from mannitol — a property possessed by few coagulase-negative strains. Mannitol-salt agar, containing an indicator to demonstrate mannitol fermentation, is a useful selective medium for identifying presumptive *Staph. aureus* in mixed cultures.

Typing:

Bacteriophage typing: strains of *Staph. aureus* can be distinguished by the pattern of their susceptibility to a set of 24 bacteriophages (phages); most strains are lysed by more than one phage and several hundred different phage types can be recognised. The phages are in three groups (I, II and III) to which the *Staph. aureus* strains can be assigned; these phage groups correspond broadly to three major serological groups identified antigenically.

Toxins:

Staph. aureus forms a large number of extracellular toxins and enzymes. Not all strains produce the whole range listed in Table 7.1; most of the products probably play a role in pathogenicity.

Staph. epidermidis produces few if any of the toxins elaborated by *Staph. aureus.*

Table 7.1 Toxins and toxic components produced by *Staphylococcus aureus*

Toxin	Activity
Haemolysins α,β,γ and δ	Cytolytic, lyse erythrocytes of various animal species
Coagulase	Clots plasma
Fibrinolysin	Digests fibrin
Leucocidin	Kills leucocytes
Hyaluronidase	Breaks down hyaluronic acid
DNAase	Hydrolyses DNA
Lipase	Lipolytic (produces opacity in egg-yolk medium)
Protein A	Antiphagocytic
Exfoliatin	Epidermal splitting
Enterotoxin(s)	Causes vomiting and diarrhoea

PATHOGENICITY

Staph. aureus, the main pathogenic species, is an important pyogenic organism: below is a list of some of the diseases it causes.

1. *Superficial infections:* pustules, boils, carbuncles, abscesses, impetigo, sycosis barbae, conjunctivitis, wound infections (including post-operative sepsis).
2. *Skin exfoliation:* toxic epidermal necrolysis (Ritter-Lyell's disease).
3. *Deep infections:* septicaemia, endocarditis, pyaemia, osteomyelitis, pneumonia.
4. *Toxic food poisoning.*

Staph. epidermidis is of much lower pathogenicity and rarely causes infections in healthy people: below are some of the diseases with which it is associated.

1. *Infective endocarditis* (especially with prosthetic heart valves).
2. *Infection of Spitz Holter valves* (in hydrocephalus).
3. *Infection of cannulae* (intra-vascular or intra-peritoneal)

Staph. saprophyticus is a major cause of urinary tract infection in sexually active women.

ANTIBIOTIC SENSITIVITY

Staph. aureus shows exceptional ability to appear in multiply-resistant form — especially in hospitals. Sensitivity testing is therefore essential in the choice of an appropriate drug for therapy.

Antibiotics active against *Staph. aureus* are:

Penicillin (50 per cent of domiciliary and 80 per cent or more of hospital strains are now resistant)

Flucloxacillin (stable to β-lactamase produced by penicillin-resistant strains and therefore effective against them)

Cephalosporins

Tetracyclines (many hospital strains resistant)

Erythromycin

Lincomycin and Clindamycin

Fusidic acid.

Penicillin resistance is due to production of *β-lactamase* which breaks down the antibiotic: the β-lactamase is plasmid-coded and transferred by transduction via bacteriophage.

Staph. epidermidis is often resistant to penicillin; most strains are sensitive to at least some of the other antistaphylococcal antibiotics.

MICROCOCCI

Micrococci are often similar to and difficult to separate from coagulase-negative staphylococci. Many strains are strict aerobes and a useful differential test is their inability to form acid from glucose under anaerobic conditions.

Colonies are usually white but those of some species are brightly pigmented yellow, orange or pink.

Micrococci have little pathogenic potential.

ANAEROBIC CLUSTERING GRAM-POSITIVE COCCI

These are best considered with the other anaerobic cocci (see Chapter 19) — classification into 'anaerobic staphylococci' and 'anaerobic streptococci' is no longer considered justifiable.

8

Streptococcus and pneumococcus

Streptococci include pneumococci (also known as *Streptococcus pneumoniae*); however, pneumococci can be considered separately as they form a distinct species with its own properties and characteristics.

STREPTOCOCCI

Classification: is extremely complex and still incomplete. An important basis for classification is the type of haemolysis produced around colonies growing on blood agar (Table 8.1).

Table 8.1 Classification of streptococci based on haemolysis

Haemolysis	Appearance	Designation	Streptococcal class
Complete	Colourless clear, sharply defined zone	β	Pyogenic streptococci
Partial	Greenish discolouration	α	Viridans streptococci
None	No change	γ or non-haemolytic	Enterococci

[handwritten annotations: Str. pyogenes A, B, C, G; Strep pyogenes — D]

Note. This classification is not absolute and variable haemolysis is produced by some strains within the groups.

Pyogenic streptococci
This class includes the most pathogenic human species: the main pathogen is *Streptococcus pyogenes*. Pyogenic streptococci have polysaccharide Lancefield group antigens: *Strep. pyogenes* is Lancefield group A; other pyogenic streptococci belong to Lancefield groups B, C and G.

Viridans streptococci

A heterogeneous group, sometimes called indifferent streptococci: strains may possess one of a variety of Lancefield group antigens (A, C, E, F, G, H, K, M or O) or none at all.

Enterococci

All enterococci have the same teichoic acid group antigen of Lancefield group D; their normal habitat is the gut.

GENERAL PROPERTIES OF STREPTOCOCCI

Morphology and staining: Gram-positive spherical or oval cocci in pairs or chains: 0.7–0.9 μm diameter.

Culture: grow well on blood agar: enrichment of media with blood, serum or glucose may be necessary:

Selective media containing aminoglycosides or 1:500 000 crystal violet inhibit other bacteria in a mixed culture but permit growth of streptococci.

Lactic acid is the major fermentative product: it accumulates in cultures and rapidly terminates growth. Culture in buffered glucose medium (e.g. Todd–Hewitt broth) increases the yield of organisms.

Aerobic: but grow well *anaerobically:* growth enhanced by 10 per cent carbon dioxide.

Colonial morphology: usually small, except for enterococci.

Haemolysis: see Table 8.1.

Serology: identifies *Lancefield groups* by group-specific polysaccharides in cell wall.

Method: classically a precipitin reaction with antigen extracted by acid treatment; now carried out by slide agglutination.

Bacitracin. Strep. pyogenes is more sensitive to bacitracin than other haemolytic streptococci. This property is the basis of a disc-diffusion sensitivity test for the presumptive identification of Lancefield group A streptococci.

The main medically important Lancefield groups with their type species are as follows:

Group A: *Strep. pyogenes*
Group B: *Strep. agalactiae*
Group C: *Strep. equisimilis*
Group D: *Strep. faecalis*
Group G: no type species identified.

LANCEFIELD GROUP A STREPTOCOCCI

Streptococcus pyogenes

The most pathogenic member of the genus: produces a large number of powerful enzymes and toxins.

Habitat: present as a commensal in the throat of a variable proportion of healthy adults and children.

LABORATORY CHARACTERISTICS

Culture: blood agar with small, typically matt or dry colonies surrounded by β haemolysis.

Toxins:

1. *Streptokinase:*
   a. Causes fibrinolysis by catalysing conversion of plasminogen to plasmin
   b. Two types produced — A and B
   c. Also produced by some strains of streptococci in Lancefield groups B, C, G and F

2. *Hyaluronidase:*
   a. Attacks hyaluronic acid — the cement of connective tissue — causing increased permeability
   b. Antibodies to this enzyme are produced after infection

3. *DNAases:*
   a. Four immunologically distinct types — A, B, C and D; B enzyme is the most common
   b. Antibodies to DNAase — especially B enzyme — are demonstrable in patients after most infections with *Strep. pyogenes*
   c. Also produced by some streptococci in Lancefield groups B, C, G and H.

4. *DPNase* (diphosphopyridine nucleotidase)
   a. Kills leucocytes
   b. Antibody formed after infection
   c. Not all strains *Strep. pyogenes* produce this enzyme: it is particularly associated with the nephritogenic strains belonging to Griffith's type 12.

5. Other enzymes produced include
   a. Leucocidin
   b. Protease
   c. Amylase

6. *Haemolysins:* two streptolysins — or toxins which lyse erythrocytes — are produced; their characteristics are shown in Table 8.2.

Table 8.2   Streptolysins of *Streptococcus pyogenes*

Haemolysin	Stability to oxygen	Synthesised anaerobically	Antigenic
Streptolysin O	–	+	+
Streptolysin S	+	–	–

+ = yes
– = no

7. *Erythrogenic toxin*
   a. Produced as a result of the presence of a lysogenic phage in the streptococci
   b. Three immunologically distinct toxins — A, B and C
   c. Responsible for the characteristic erythematous rash in scarlet fever
   d. Dick test. Injection intradermally produces localised area of erythema within 8–16 h in non-immune people; presence of antibody to the toxin results in neutralisation of the toxin and no reaction is produced

   Note. All these enzymes and toxins probably contribute to the invasiveness and pathogenicity of *Strep. pyogenes.*

Serotypes
Strep. pyogenes (Lancefield group A): can be subdivided into Griffith types depending on three surface protein antigens:

   M: Type-specific, i.e. there is a distinct M antigen for each type or strain; only found in virulent or pathogenic strains; 65 distinct M serotypes have been identified.
   R: Not related to virulence; fewer antigens and the same R antigen can be found in several different M types. Also found in a few strains in Lancefield groups B, C and G
   T: There are also T antigens each one of which may be found on several different M types; used in conjunction with M typing for identifying different types of *Strep. pyogenes*

PATHOGENICITY
Streptococcus pyogenes causes:
   Tonsillitis and pharyngitis
   Peritonsillar abscess (quinsy)
   Scarlet fever
   Otitis media
   Mastoiditis
   Puerperal sepsis
   Impetigo

Post-streptococcal complications:
Rheumatic fever
Glomerulonephritis
* Erythema nodosum

ANTIBIOTIC SENSITIVITY

Penicillin: drug of choice.
   In patients hypersensitive to penicillin: erythromycin.
   Antibiotic resistance: is rare — penicillin resistance is unknown but resistance to tetracycline is not uncommon.

LANCEFIELD GROUP B STREPTOCOCCI

Group B contains only one species: Strep. agalactiae — increasingly recognised as an important human pathogen.
   Habitat: commensal of female genital tract: common in animals — especially cattle — in which it causes bovine mastitis.
   Culture: grows on ordinary and bile-containing (MacConkey) media.
   Colonies: may be α, β or non-haemolytic: typically produces red or orange pigment when incubated anaerobically on starch-containing media.
   Identification: Lancefield grouping.
   Pathogenicity: an important pathogen in neonates causing meningitis and septicaemia: also associated with septic abortion, puerperal or gynaecological sepsis.
   Antibiotic sensitivity: penicillin, erythromycin.

LANCEFIELD GROUP C STREPTOCOCCI

Rarely cause human disease.
   Culture: blood agar; colonies often large and mucoid due to hyaluronic acid capsules. Usually β haemolytic.
   Identification: Lancefield grouping.
   Biotypes: four species can be distinguished by biochemical and other tests:

Strep. equisimilis	most common type isolated from humans: produces streptokinase
Strep. equi ⎫	rarely
Strep. dysgalactiae ⎬	found
Strep. zooepidemicus ⎭	in humans

Note. Some strains of *Strep. milleri* (see under viridans streptococci) have group C antigen.

Pathogenicity. Low pathogenicity for man: occasionally cause of tonsillitis; rarely endocarditis, glomerulonephritis.

Antibiotic sensitivity: penicillin.

LANCEFIELD GROUP D STREPTOCOCCI

This group includes the enterococci — important human pathogens second only to *Strep. pyogenes* in pathogenicity.

Morphology: oval cocci, usually in pairs, do not readily chain.

Culture: grow on ordinary and bile-containing media: heat-resistant and able to grow at 45°C; also able to grow in the presence of 6.5 per cent sodium chloride.

Colonies: large, whiteish, haemolysis variable but generally non-haemolytic; lactose-fermenting colonies on MacConkey agar (pink) and on CLED agar (yellow).

Identification: colonial and cultural characteristics; demonstration of Lancefield group D antigen.

Species: the medically important species are listed in Table 8.3.

Table 8.3   Group D streptococci

Species	Habitat	Disease
Strep. faecalis	Human gut	Urinary, abdominal wound infection: rarely endocarditis
Strep. faecium *Strep. durans*	Rarely human gut, faeces of domestic animals	As above
Strep. bovis (sometimes classified as a viridans streptococcus)	Animal gut	Endocarditis

Antibiotic sensitivity: enterococci are sensitive to ampicillin, moderately resistant to penicillin and resistant to the cephalosporins.

LANCEFIELD GROUP G STREPTOCOCCI

A heterogenous group: information on the streptococci within it is scanty and no species have so far been recognised.

Habitat: human throat

Culture: two colonial forms both producing β haemolysis

1. Large colony: strains resemble *Strep. pyogenes*: pathogenicity low, tonsillitis, rarely endocarditis.

2. Small colony: strains are occasionally found in the human throat, possibly gut: low pathogenicity — rarely abdominal sepsis, tonsillitis.

Note. Some strains of *Strep. milleri* possess Group G antigen.

VIRIDANS OR INDIFFERENT STREPTOCOCCI

An ill-defined group which typically show α haemolysis on blood agar: but haemolysis is variable and some strains are non-haemolytic: most human strains are commensals of the upper respiratory tract and of low pathogenicity.

Clinical laboratories usually do not differentiate species but simply report *Strep. viridans*: species identification within the viridans group depends on the results of a range of biochemical tests: species do not possess a characterising Lancefield group antigen.

The principal species are listed in Table 8.4.

Table 8.4   Viridans (indifferent) streptococci

Species	Haemolysis on blood agar	Lancefield group antigens	Habitat	Disease
Strep. mitior	α	O, K, M, Q or none		Endocarditis
Strep. sanguis	α	H, K or none		Endocarditis
Strep. mutans	None	A or none	Human upper respiratory tract	Endocarditis, dental caries
Strep. milleri	None	A, C, F, G or none		Deep abdominal, brain, abscesses
Strep. salivarius	None	K or none		Rarely, endocarditis

Note: Strep. milleri is now recognised as an important cause of sepsis.

Streptococcus pneumoniae

Probably the most common and nowadays the most important pathogen amongst the streptococci:

Also known: as pneumococcus or *Diplococcus pneumoniae.*

Habitat: normal commensal of the upper respiratory tract.

LABORATORY CHARACTERISTICS

Morphology: lanceolate diplococci arranged longitudinally in pairs with the pointed ends outwards: sometimes forms short chains: normally capsulated with carbohydrate antigenic capsule which is correlated with virulence (Fig. 8.1).

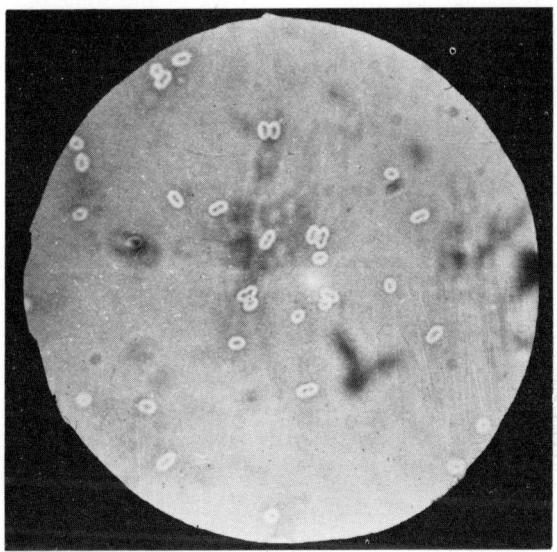

Fig. 8.1   Pneumococci. Stained to show capsules (magnified × 2000).

Culture: blood agar, sometimes broth enriched with serum or glucose.

Colonies: α-haemolytic, typically 'draughtsman', i.e. with sunken centre due to spontaneous autolysis of older organisms: young colonies may resemble dew drops due to large capsules before autolysis:

Identify: differentiate from viridans streptococci by:-

1. *Optochin sensitivity:* viridans streptococci are resistant.
2. *Bile solubility:* addition of bile to broth cultures lyses pneumococci but not viridans streptococci.
3. *Inulin fermentation:* viridans streptococci do not ferment inulin.

Antigenic structure:

1. *Capsule:* consists of the C carbohydrate antigen: type specific: 83 capsular types are recognised.
   Identified by capsule swelling — quellung reaction — observed microscopically when pneumococci are mixed with specific antisera.
2. *Somatic antigen:* common to all pneumococci: possibly choline teichoic acid.
3. *Protein M antigen:* unrelated to M antigens of *Strep. pyogenes.*

Virulence correlates with the presence of a capsule probably because this prevents or inhibits phagocytosis.

Transformation: the transfer of DNA — and some of the genetic markers for which it codes — from one bacterial strain to another was first demonstrated in pneumococci by Griffiths in 1928.

Pneumococcal types: not all types are equally common and the majority (83 per cent) of human infections are associated with 15 types.

Common infecting types in Britain in descending frequency are: types 3, 14, 1, 19, 8, 18, 6, 23, 7, 4, 9, 5, 12, 22 and 11.

The main types associated with pneumonia are types 1, 3 and 8: meningitis, on the other hand, is more often due to types 6, 18, 19 and 23.

PATHOGENICITY

Pneumococci are important pathogens and cause a considerable amount of both morbidity and mortality today despite their sensitivity to penicillin.

Lobar pneumonia
Acute exacerbation of chronic bronchitis (often with *H. influenzae*)
Meningitis
Otitis media
Sinusitis
Conjunctivitis
Septicaemia (especially in splenectomized patients)

ANTIBIOTIC SENSITIVITY

Sensitive to penicillin, erythromycin, cotrimoxazole, tetracycline.

9

Enterobacteria

Gram-negative bacilli which belong to the tribe Enterobacteriaceae. Often called 'coliform bacilli', or coliforms. Intestinal parasites of man and animals, many are human pathogens.

Table 9.1 lists the main medically important species.

Table 9.1 Medically important enterobacteria

Genus	Species	Principal diseases
Escherichia	*Esch. coli*	Wound and urinary infection; gastroenteritis in children
Shigella	*Sh. dysenteriae* *Sh. flexneri* *Sh. boydii* *Sh. sonnei*	Dysentery
Salmonella	*S. typhi* *S. paratyphi* A, B, C *S. typhimurium* Many other serotypes	Enteric fever Food-poisoning
Klebsiella	*K. aerogenes* *K. pneumoniae*	Urinary infections, other forms of sepsis
Proteus	*Pr. mirabilis* *Pr. vulgaris* *Pr. morganii* *Pr. rettgeri*	Urinary infections, other forms of sepsis
Yersinia	*Y. pestis* *Y. pseudotuberculosis* *Y. enterocolitica* }	Plague Septicaemia, enteritis, mesenteric adenitis
Enterobacter Serratia Providencia Citrobacter	*Ent. cloacae* *Ent. aerogenes* *Serr. marcescens* *Prov. stuartii* *Prov. alcalifaciens* *C. freundii*	 Generally of lower pathogenicity; urinary infections, other forms of sepsis

LABORATORY CHARACTERISTICS

Morphology and staining. Gram-negative bacilli 1.4 x 0.6 μm: non-motile or motile by peritrichous flagella (Fig. 9.1): non sporing.

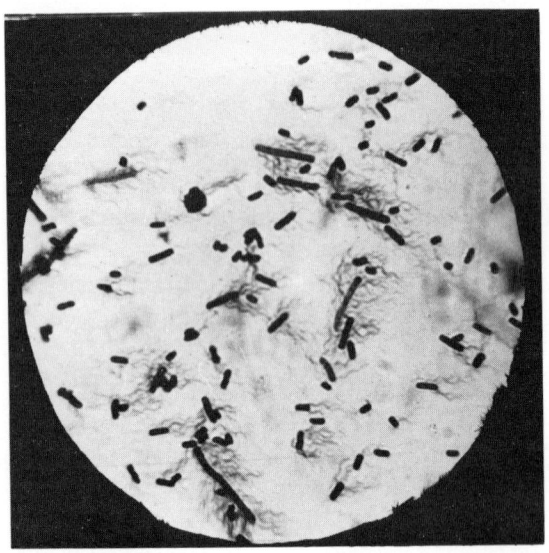

Fig. 9.1 Flagella. Peritrichous flagella demonstrated by a silver-impregnation staining method (magnified × 2000).

Culture. Grow well on ordinary media, e.g. blood agar, MacConkey agar, CLED agar; aerobic and facultatively anaerobic; wide temperature range.

Identification:

1. *Colonial morphology* on indicator media assists in initial identification: e.g. MacConkey and CLED agar contain lactose and a pH indicator so that lactose-fermenting colonies can be identified by a change in colour.

2. *Biochemical tests:* some are characteristic of all enterobacteria, e.g.
   a. Reduce nitrate to nitrite
   b. Ferment glucose with acid (and sometimes gas) production
   c. Oxidase negative

   Others are required to identify different species: formerly involving numerous individual sugar fermentation and other reactions, this is now often done by test kits.

3. *Serological tests* based on identification of somatic and flagellar antigens are used for the final identification of *Salmonella* and *Shigella species* and, in some cases, for strain identification. Strain identification within a species can also be done by

   a. Bacteriophage typing

   b. Bacteriocin typing.

Antigenic structure is complex, with considerable sharing of antigens both between species and within a species: antisera prepared in animals, and usually requiring numerous absorptions to reduce cross-reacting antibodies, are used to identify an organism on the basis of the specific antigens it possesses. A diagram of the sites of the main antigens in enterobacteria is shown in Figure 9.2.

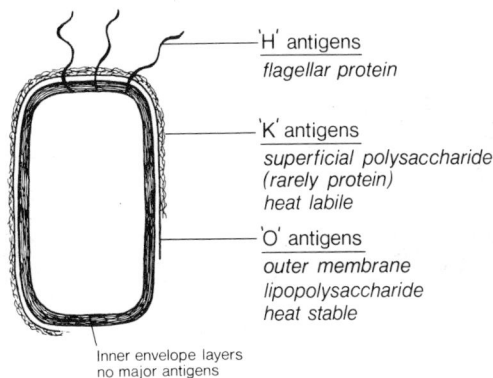

Fig. 9.2   Antigenic structure of Gram-negative bacteria.

Endotoxins are lipopolysaccharides present in the cell wall of Gram-negative bacilli (see Chapter 2): they are liberated when the bacterial cells lyse and are responsible for many pathological effects during human infection with the organisms.

Exotoxins are *proteins* liberated extracellularly from the intact bacterium by a few species of enterobacteria, e.g. *Shigella dysenteriae* produces a neurotoxin, toxigenic strains of *Escherichia coli* produce enterotoxins.

ANTIBIOTIC SENSITIVITY

Antibiotic sensitivity is variable and difficult to predict for any individual species: this is partly due to the frequency with which strains of enterobacteria acquire plasmids which may spread to other strains: plasmids may contain genes which code for resistance to several antibiotics.

Antibiotics which may be active against enterobacteria are listed below:

Ampicillin, amoxycillin and mezlocillin
Aminoglycosides
Trimethoprim
Sulphonamides
Chloramphenicol
Cephalosporins (particularly of the second and third generation)

Urinary antimicrobial drugs: in addition to those listed above, some other antibacterial drugs can be used, but only for urinary tract infections, e.g.
Nitrofurantoin
Nalidixic acid

Escherichia coli

Habitat: a normal inhabitant of the human and animal intestine.
Isolation: grows well as large colonies after overnight incubation.
Identification. Most strains are motile: some strains are capsulate. Usually ferments lactose (hence produces colonies pink on MacConkey agar, yellow on CLED agar). Indole is produced. Further biochemical tests are required for formal identification.

Typing:

1. Determination of O, K and H antigens.
2. Colicin typing: colicins are bacteriocins produced by *Esch. coli.*

Pathogenicity:

1. Urinary tract infection
2. Diarrhoea
   a.   Certain serotypes cause disease in infants and young children: explosive outbreaks in nurseries.
   b.   Some strains cause diarrhoea in other age groups.
3. Wound infections — especially after lower intestinal tract surgery (the flora of these infections often includes anaerobes).
4. Peritonitis.
5. Biliary tract infections.
6. Septicaemia.
7. Neonatal meningitis.

Shigella

Habitat: intestinal parasites of man and primates
Isolation: shigellae grow well on routine media. They do not ferment

lactose and so produce colourless colonies on MacConkey agar and DCA (a medium with lactose which also contains desoxycholate citrate to inhibit growth of *Esch. coli*).

Shigella sonnei is an important exception, it ferments lactose slowly with production of pale pink colonies.

Enrichment culture of faeces in selenite F broth, with subsequent subculture onto MacConkey agar may improve isolation.

Identification. There are four species and numerous serotypes: *Sh. dysenteriae*, 10 serotypes;

Sh. boydii, 15 serotypes;
Sh. flexneri, six serotypes;
Sh. sonnei, one serotype.

1. *Motility:* shigellae are non-motile.
2. *Biochemical tests:* shigellae produce acid but not gas from carbohydrates.
3. *Determination of O antigens.*

Typing: 'Colicin' typing is available for *Sh. sonnei*.
Pathogenicity:

Sh. dysenteriae, which produces protein exotoxins, causes the most severe illness, 'Shiga dysentery'. Dysentery due to other shigellae tends to be milder: *Sh. sonnei* is the cause of most dysentery in Britain.

Salmonella

Habitat. Animal gut: predominantly animal pathogens which can also cause disease in man. Foodstuffs from animal sources are important vehicles in the transmission of infection. *Salmonella typhi* and *S. paratyphi* differ from the other species in that man is the only natural host.

Isolation: grow well on ordinary media.

From faeces, culture on:

1. *MacConkey agar.*
2. *Desoxycholate citrate agar.*
   Observe for pale (non-lactose fermenting) colonies.
3. *Wilson and Blair's bismuth sulphite agar:* observe for black metallic colonies (due to hydrogen sulphide production).
4. *Selenite F and tetrathionate broths* for enrichment: then subculture to MacConkey agar.

Identification:

1. *Motility:* salmonellae are motile
2. *Biochemical tests:* salmonellae generally produce acid and gas from carbohydrates — except for *S. typhi* which does not produce gas.

3. *Serology:* by identification of antigens:
   O: somatic
   H: flagellar
   Vi: a surface antigen possessed by a few species including *S. typhi*

More than 1500 serotypes ('species') are recognised: sharing of O and H antigens is common and identification is complex depending on detection of several antigens (see Table 9.2).

Table 9.2   Selected salmonellae showing antigenic profiles (Kauffmann-White scheme)

Serotype ('species')	Group	O Antigens	H Antigens Phase 1	Phase 2
S. paratyphi A	A	1.2.12.	a	—
S. paratyphi B	B	1.4.5.12	b	1.2
S. agona	B	4.12.	f.g.s.	—
S. typhimurium	B	1.4.5.12	i	1.2
*S. paratyphi C**	C	6,7	c	1.5
*S. typhi**	D	9.12.	d	—

* Also possess Vi antigen.

Phase variation: a single strain can possess two different sets of H antigens at different times; both sets must normally be identified.

Typing: Bacteriophage typing of particular serotypes can be carried out e.g. Vi-phages of *S. typhi*.

Pathogenicity:

1. *Enteric fever* is due to *S. typhi, S. paratyphi A, B, C*
2. Most other salmonella serotypes (*S. typhimurium* is the most common) cause gastroenteritis or food-poisoning; however, some particular types (e.g. *S. dublin, S. cholerae-suis*) have a tendency to cause septicaemia.
3. Rarely, salmonellae cause osteomyelitis, septic arthritis, and other purulent lesions.

Klebsiella

Habitat. Human and animal intestine. Some strains are saprophytes in soil, water and vegetation. They survive well in moist environments in hospitals.

Isolation. Grow well, producing colonies which are large and mucoid (due to the possession of a prominent capsule) on blood agar, CLED agar and MacConkey agar.

Identification: Not motile. Biochemical tests: lactose is fermented. Nomenclature is confused, and there is disagreement as to the names and number of species in the genus; *Klebsiella aerogenes, K. pneumoniae, K. ozaenae, K. oxytoca* are terms commonly used.

Typing:
1. Antigenic analysis of the capsular polysaccarides which comprise the large capsules typical of this genus. More than 80 serotypes recognised.
2. Klebicin (bacteriocin) typing.

Pathogenicity:
1. Urinary tract infection.
2. Septicaemia.
3. Meningitis (especially in neonates).
4. Rarely abscesses, endocarditis and other lesions. Sometimes associated with chronic nasal and oropharyngeal sepsis.
5. Friedlander's pneumonia.

Proteus
Habitat. Human and animal intestine. Some strains are saprophytes and are found in soil and water.

Isolation: grow well on routine media: a swarming type of growth which may cover the whole plate is produced on ordinary media by *Proteus mirabilis* and *Pr. vulgaris:* discrete colonies are produced by *Pr. morganii* and *Pr. rettgeri:* proteus forms colourless colonies on MacConkey agar and blue-green colonies on CLED agar.

Identification: Swarming growth allows presumptive identification of *Pr. mirabilis* or *Pr. vulgaris.* Biochemical tests permit formal identification. Lactose is not fermented. *Pr. mirabilis* is indole-negative, the other species are indole-positive. Proteus species produce a potent urease.

Typing:
1. Serotyping.
2. Bacteriocin typing.
3. 'Dienes' phenomenon — inhibition of swarming by dissimilar strains of *Pr. mirabilis.*

Pathogenicity:
Pr. mirabilis is the most frequently isolated species.

1. Urinary tract infection: urinary urea 'split' by urease to produce ammonium salts (alkaline pH); association with urinary calçuli.

2. Often isolated from mixed flora of wounds, burns, pressure sores, chronic discharging ears — generally a low-grade pathogen in such circumstances but occasionally these sites provide a portal of entry for septicaemia or brain abscess (ear infections).
3. Septicaemia.

Yersinia

Habitat. Yersinia are animal parasites which sometimes — although rarely — cause disease in man.

Isolation: from blood or tissue (when samples are available); difficult from mixed flora of faeces.

Colonies are often smaller than those of other enterobacteria, and prolonged incubation of cultures is therefore necessary.

Identification. Generally small (2.0 x 0.5 μm) bacilli which may show 'bi-polar' staining. *Yersinia pestis* is non-motile; *Y. pseudotuberculosis* and *Y. enterocolitica* are motile at 22°C. Biochemical tests permit formal identification, the reactions being more reproducible at 22–27°C than the 37°C temperature used for other enterobacteria.

Identification of *Y. pestis* is simplified by fluorescent-antibody staining and by testing for susceptibility to specific bacteriophage.

Pathogenicity:

1. *Plague: Y. pestis* (formerly called *Pasteurella pestis*) causes bubonic and pneumonic plague ('The Black Death').
2. *Y. pseudotuberculosis* and *Y. enterocolitica* cause.
   a. A septicaemic illness similar to typhoid fever
   b. Enteritis
   c. Mesenteric adenitis, sometimes associated with terminal ileitis. Clinically, this closely mimics appendicitis

Enterobacter; Serratia; Providencia; Citrobacter

These members of the enterobacteria can conveniently be considered together.

Habitat. Human and animal intestinal parasites, but some strains are saprophytes. Moist environments in hospitals may be important reservoirs.

Isolation. They grow well on routine media.

Identification: biochemical tests.

The most frequently isolated species are:

Enterobacter cloacae, Ent. aerogenes, Serratia marcescens, Providencia stuartii, Citrobacter freundii.

Pigment: Some strains of *Serr. marcescens* produce characteristic magenta-coloured colonies.

Pathogenicity

1. Urinary tract infections (particularly chronic, complicated infections).
2. Wounds, skin lesions and respiratory infections in hospitalised patients.
3. Septicaemia.

They have been responsible for outbreaks of infection in intensive care areas, burns units and other special units; they are often resistant to many antibiotics.

10

Pseudomonas and other aerobic Gram-negative bacilli

PSEUDOMONAS

Gram-negative motile aerobic bacilli having very simple growth requirements: widely distributed in water, soil and sewage: the genus includes pathogens for animals and plants.

Pseudomonas aeruginosa (Pseudomonas pyocyanea)
Habitat: human and animal gastrointestinal tract, water, soil. Moist environments are important reservoirs in hospitals and *Pseudomonas aeruginosa* is able to survive and multiply in certain aqueous antiseptics and other fluids.

LABORATORY CHARACTERISTICS

Morphology and staining: Gram-negative bacilli (1.5 – 3.0 μm × 0.5 μm), motile (polar flagellate), non-sporing, non-capsulate.

Culture: a strict aerobe; grows readily on all routine media over a wide temperature range (5°C to 42°C).

Broth cultures: show uniform turbidity and a surface pellicle — evidence of its aerobic growth requirement.

Plate cultures: some colonies are 'coliform-like' but others typically large and irregular; they have a fluorescent, greenish appearance due to production of *pyocyanin* (blue-green) and *fluorescein* (yellow) pigments and a characteristic 'fruity' odour. Strains isolated from sputum of patients with cystic fibrosis often give rise to large mucoid colonies due to formation of extracellular polysaccharide slime.

Selective media:

1. *MacConkey agar:* yellowish non-lactose fermenting colonies.
2. *Cetrimide agar:* this medium inhibits many other organisms but allows *Ps. aeruginosa*, which resists the action of quaternary ammonium compounds, to grow.

Identify by:

Colonial morphology.

Biochemical tests: acid production in a range of carbohydrates which are oxidised not 'fermented': oxidase-positive (this is a valuable differential test); liquifies gelatin.

Examination of cultures under ultraviolet light: colonies display intense greenish fluorescence — useful to detect *Ps. aeruginosa* in mixed cultures.

Typing:

Pyocin (bacteriocin) typing: a method that can be carried out in routine diagnostic laboratories. The sensitivity of eight indicator strains of *Ps. aeruginosa* to the pyocins produced by the test strain is noted and from the pattern of growth inhibition about 40 types can be recognised.

Phage typing: considerable difficulty has been encountered in obtaining reproducable results.

Antigenic structure: a typing scheme exists based on the heat-resistant somatic ('O') antigens.

Useful in epidemiological investigations.

PATHOGENICITY

An important cause of hospital-acquired infections. Most severe infections occur in patients with serious underlying conditions (e.g. burns, malignancy), or as a result of therapeutic procedures (e.g. indwelling urinary tract catheters, mechanical ventilatory support). Previous antibiotic therapy also favours infection with *Ps. aeruginosa*.

1. Urinary tract infection:
   a. chronic, complicated infections
   b. association with indwelling catheter.
2. Burns.
3. Septicaemia — 'ecthyma gangrenosum' skin lesions may be present.
4. Wound infections.
5. Infected skin lesions, e.g. pressure sores, varicose ulcers.
6. Chronic otitis media and externa.
7. Lower respiratory tract infections
   a. in cystic fibrosis
   b. in patients on ventilators.
8. Eye infections, secondary to trauma or surgery. Loss of sight may result.

Vaccine

Sixteen serotypes of *Ps. aeruginosa* are now internationally recognised and a polyvalent vaccine made from the cell-surface components of representative strains has been developed. This vaccine, although not yet commercially available, stimulates active immunity in man (and protective human immunoglobulin for passive immunisation has been produced from the serum of vaccinated volunteers). It may be of value in the prevention of pseudomonas infections in susceptible patients: the efficacy of vaccination (active and passive) has been demonstrated in patients with severe burns in field trials conducted in India.

ANTIBIOTIC SENSITIVITY

Resistant to most usual antibiotics. The major antipseudomonal drugs are:

Aminoglycosides

Certain penicillins (carbenicillin, ticarcillin, azlocillin).

Polymyxins B and E (colistin).

Note. A laboratory report of *Ps. aeruginosa*, particularly if isolated from a superficial site, is not an indication for prescription of toxic and expensive antibiotics. A careful clinical and bacteriological assessment of the likely relevance of the information in the individual patient should always be made.

OTHER PSEUDOMONAS SPECIES

Pseudomonas fluorescens and pseudomonas putida

These fluorescent pseudomonads are similar to *Pseudomonas aeruginosa*, they differ in their failure to produce pyocyanin and in some biochemical reactions. They are often regarded as biotypes of *Ps. aeruginosa*.

Many strains of *Ps. fluorescens* and *Ps. putida* grow slowly at 4°C and are sometimes isolated from contaminated blood which has been stored at this temperature: such blood, if transfused, causes severe reactions.

Other non-fluorescent species of Pseudomonas which lead a primarily saprophytic existence are recognized. Compared to *Ps. aeruginosa*, their ability to cause human infection is limited. However, such organisms (e.g. *Ps. cepacia*, *Ps. maltophilia*) do occasionally cause serious human infection, particularly when host defences are impaired.

Pseudomonas mallei and pseudomonas pseudomallei

These antigenically related bacteria cause similar diseases. They are now recognised to be non-fluorescent pseudomonads but in the past they were classified in genera that have been discarded, e.g. *Pfeifferella, Loefflerella, Malleomyces.*

Ps. pseudomallei grows readily but *Ps. mallei*, unlike other pseudomonas, grows poorly on ordinary laboratory media, is non-motile and is a strict animal parasite.

Ps. pseudomallei (Whitmore's bacillus) causes melioidosis, a disease of animals and humans, endemic in South-East Asia. Most human cases are asymptomatic. Pulmonary consolidation, skin lesions and a fatal septicaemia may occur. The organism is a saprophyte of certain soils and waters and there may be a large animal reservoir locally.

Ps. mallei causes glanders in horses. Rarely, human infections occur, acquired from animals or from laboratory work with the organism.

OTHER AEROBIC GRAM-NEGATIVE BACILLI

Acinetobacter

.Gram-negative cocco-bacillus (microscopic morphology may be confused with Neisseria). Strictly aerobic non-motile, oxidase-negative. Species include *Acinetobacter calcoaceticus.*

Habitat: widely distributed in nature. Human skin carriage may be an important cause of dissemination in hospitals.

Isolation: grows well on routine media.

Identification: biochemical tests.

Pathogenicity: generally a low-grade pathogen: increasingly being isolated from hospitalised patients, particularly from those with burns and in intensive care units. May colonise intravascular cannulae.

Antibiotic sensitivity: variable: often resistant to many antibiotics, including those commonly used to treat other forms of serious hospital sepsis.

Moraxella

Gram-negative cocco-bacillus; strictly aerobic; non-motile, oxidase-positive; sensitive to penicillin. *Moraxella lacunata* causes a purulent conjunctivitis.

Aeromonas: Plesiomonas

Gram-negative bacilli; facultative anaerobes: motile; oxidase-positive.

Aeromonas hydrophila and *Plesiomonas shigelloides* are the medically important species but infections are rare: occasionally isolated from blood, CSF and exudates. Outbreaks of gastroenteritis due to *P. shigelloides* have been reported in the Far East. Sporadic cases are occasionally seen in Britain.

Streptobacillus

Streptobacillus moniliformis is a slender branching filamentous bacterium with club-shaped ('moniliform') terminal swellings. Gram-negative but may be Gram-positive in young cultures. Requires enriched media for growth. Causes one form of rat-bite fever in man.

11

Vibrio and campylobacter

Actively motile, Gram-negative curved bacilli.

VIBRIO

Widespread in nature, mainly in water: one species, *Vibrio cholerae*, is the cause of *cholera*.

Vibrio cholerae (Vibrio comma)
Habitat: water contaminated with faeces of patients or carriers.

LABORATORY CHARACTERISTICS

Morphology and staining. Gram-negative slender bacilli (2 × 0.5 μm), sometimes comma-shaped with a pointed end (Fig. 11.1); often arranged in pairs or short chains giving a spiral appearance. Actively motile by one long polar flagellum; non-capsulate, non-sporing.

Culture. Aerobe; grows readily on ordinary media as glistening colonies over a wide temperature range (optimum 37°C). Growth is inhibited at acid pH (optimal pH for growth is alkaline from 8.0–8.2).

Enrichment medium. Alkaline peptone water (pH 8.4) promotes the rapid growth of *V. cholerae* from mixtures of other bacteria, e.g. in faecal samples.

Selective media:

1. Dieudonné's medium — alkaline blood agar.
2. Aronson's medium — alkaline sucrose dextrin agar
   Nowadays these media have been largely replaced by
3. Monsur's medium — gelatin taurocholate tellurite agar pH 8.5–9.2
   Observe: for large colonies, grey at 24 h, with a black centre at 48 h, surrounded by a halo due to gelatin liquefaction.
4. TCBS medium — thiosulphate citrate bile sucrose agar pH 8.6
   Observe: for large yellow sucrose-fermenting colonies after 18–24 h incubation. Enterobacteria may grow but growth is inhibited and the colonies are small.

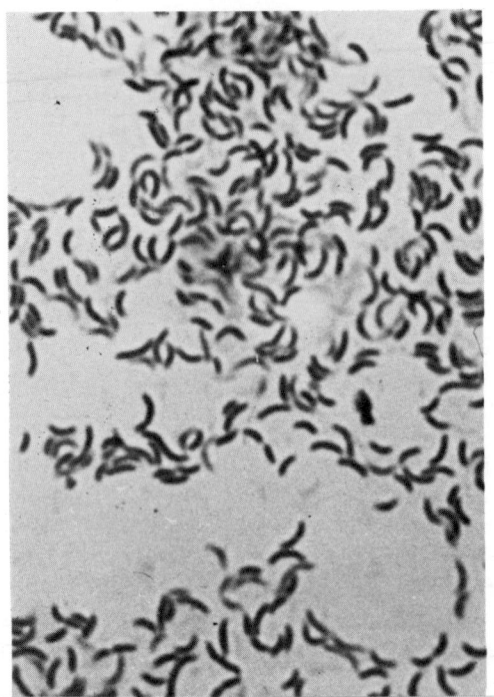

Fig. 11.1   Vibrios. The characteristic comma-shape of the curved bacilli can be seen (magnified × 4000).

Identify by slide agglutination of suspect colonies with specific antisera.

Biochemical reactions. Fermentation of sucrose and mannose (acid, no gas) but not arabinose is typical of *V. cholerae* and a distinguishing feature from other vibrios.

Cholera red reaction: add a few drops of sulphuric acid to a 24 h peptone-water culture.

Observe: for immediate formation of red-pink colour (this reaction is given by many vibrios and is not specific for *V. cholerae*).

Mechanism: production of indole and nitrites in the peptone water results in formation of the coloured substance nitroso-indole.

Biotypes. There are two biotypes of *V. cholerae*: (1) classical and (2) El Tor — differentiated as shown in Table 11.1. (Biotyping is the distinguishing of different bacterial strains within a species by various biological and biochemical reactions).

Antigenic structure:

H antigens are common to many vibrio species but of little value in identification.

Table 11.1 Biotypes of *Vibrio cholerae*

Test	*V. cholerae* (classical)	*V. cholerae* (El Tor)
Action of cholera group iv phage	Sensitive (lysis)	Resistant (no lysis)
Action of cholera group v phage	Resistant (no lysis)	Sensitive (lysis)
Soluble haemolysin production	—	+
Haemagglutination of fowl cells	—	+
Voges-Proskauer reaction	—	+
Action of polymyxin (50 units)	Sensitive	Resistant

O antigens enable the vibrios to be divided into six major subgroups. All strains of *V. cholerae* possess a distinctive O antigen and belong to subgroup I. Subsidiary O antigens allow subdivision of *V. cholerae* into three serotypes, *Ogawa, Inaba* and *Hikojima*; however, antigenic structure (and, therefore, serotype) may change within the human gut. Any serotype can be either classical or El Tor biotype.

Vibrios deficient in the O antigen of subgroup I are classified as non-cholera or non-agglutinable vibrios (NCV or NAG).

Phage typing. Five phage types of classical *V. cholerae* and six phage types of *V. cholerae* El Tor have been described. Phage typing is of only limited value in epidemiological studies.

Toxins. Endotoxins (cell-wall lipopolysaccharide) and exotoxins are recognised. The enterotoxin is an exotoxin which stimulates persistent and excessive secretion of isotonic fluid by the intestinal mucosa.

PATHOGENICITY

Vibrio cholerae is the cause of cholera in man. In the acute disease vibrios are present in enormous numbers — about 10^8 per ml of faeces; some animals are susceptible to laboratory challenge and can serve as models for the study of the disease.

Viability. Readily killed by heat and drying; dies in polluted waters but may survive in clean stagnant water (especially if alkaline) or sea water for 1 to 2 weeks. El Tor biotype is more resistant to adverse conditions than classical *V. cholerae*.

ANTIBIOTIC SENSITIVITY

Sensitive to a wide range of antibiotics including tetracycline and chloramphenicol.

Beneckea

V. parahaemolyticus (now reclassified as *Beneckea parahaemolyticus*) is a halophilic marine vibrio isolated from shellfish particularly in countries with warm coastal waters, e.g. South-East Asia.

It causes an acute gastroenteritis in which vibrios are excreted in large numbers in the stools. Faecal samples plated on TCBS agar yield large blue-green colonies typical of *V. parahaemolyticus*.

CAMPYLOBACTER

Strictly microaerophilic vibrios have been placed in a new genus — *Campylobacter*.

Campylobacter fetus causes infertility and abortion in cattle. It grows well at 25°C but not at 43°C.

C. coli and C. jejuni. These species are difficult to differentiate and are therefore conveniently designated *C. coli/jejuni*. Recently recognised as a major cause of diarrhoeal disease in man. Thermophilic, i.e. grow well at 43°C but not at 25°C.

Campylobacter coli/jejuni

Habitat: gastrointestinal tract of farm animals (poultry, cattle, sheep, pigs) and domestic pets (dogs — especially puppies — and cats); often present in sewage.

LABORATORY CHARACTERISTICS

Morphology and staining. Small, Gram-negative, curved or spiral rods; highly motile by a single flagellum at one or both poles.

Culture. Microaerophilic; grow best in an atmosphere containing a mixture of 7 per cent oxygen, 10–15 per cent carbon dioxide with the remainder an inert gas, usually nitrogen or hydrogen.

Growth takes place at 37°C but the optimal temperature is 43°C.

Media. Grow readily on simple media.

Selective medium (necessary for isolation from faeces and other samples containing numerous other bacteria): lysed blood agar with vancomycin, polymyxin and trimethoprim.

Incubate: for 24–48 h at 43°C under microaerophilic conditions.

Observe: for effuse colonies that look like spreading fluid droplets.

Identify by Gram-film appearance, motility, growth temperature requirements (25°C—, 37°C+, 43°C++) and oxidase test (campylobacters are oxidase-positive).

PATHOGENICITY

Responsible for febrile diarrhoeal illness in man: occasionally cause septicaemia.

ANTIBIOTIC SENSITIVITY

Sensitive to macrolide (e.g. erythromycin) and aminoglycoside antibiotics, partially sensitive to ampicillin, resistant to penicillin.

12

Parvobacteria

Parvobacteria (*parvus* = small) are a heterogeneous group which contains several important human pathogens causing a wide variety of different diseases. Parvobacteria are small, Gram-negative bacilli which generally require enriched media for isolation and culture.
  Parvobacteria contain the following genera:

Haemophilus
Brucella
Bordetella
Pasteurella
Francisella
Actinobacillus.

HAEMOPHILUS

Habitat: mainly the respiratory tract: often part of the normal flora but may also cause respiratory disease.

LABORATORY CHARACTERISTICS

Morphology and staining. Small Gram-negative coccobacilli, non-sporing, non-motile.
  Culture: requires enriched media such as blood or chocolate agar and an atmosphere with added CO_2; enriched media are necessary because Haemophilus species need one or both of two growth factors:
  X factor — haematin
  V factor — diphosphopyridine nucleotide.
  Requirement for growth factors can help to differentiate species (Table 12.1).

Table 12.1 Growth factors for *Haemophilus species*

Factor required	Species
X and V	H. influenzae
X	H. ducreyi
V	H. parainfluenzae

PATHOGENICITY

The diseases caused by *Haemophilus* species are listed in Table 12.2.

Table 12.2 Pathogenicity of *Haemophilus species*

Species	Disease
H. influenzae	Exacerbations of chronic bronchitis Meningitis Epiglottitis Sinusitis, otitis media
H. aegyptius	Conjunctivitis
H. ducreyi	Chancroid
H. parainfluenzae *H. haemolyticus* *H. parahaemolyticus* }	Commensals of the upper respiratory tract; rarely cause disease

Haemophilus influenzae

The main pathogenic species.

Habitat: the upper respiratory tract: most strains found in the normal flora are non-capsulated.

Laboratory characteristics

Morphology: strains isolated from acute infections are usually capsulated.

Culture: on chocolate or blood agar: a streak of *Staphylococcus aureus* across the plate produces V factor and enlarges the size of adjacent colonies of *H. influenzae*.

Colonial morphology: small translucent, non-haemolytic colonies: enlarged if growing adjacent to a streak of *Staph. aureus* — *satellitism* (Fig. 12.1).

Identify: by testing on nutrient agar for requirements for X and V growth factors.

Antigenic structure:

Serotypes. Six are recognized — on the basis of capsular polysaccharide antigens — Pittman types a, b, c, d, e, f. Type b is the main cause of haemophilus meningitis and acute epiglottitis.

Pathogenicity

Non-capsulated strains are mainly responsible for exacerbations of chronic bronchitis

Capsulated strains (predominately type b) can cause various infections mainly in children aged from 2 months to 3 years:
Meningitis
Acute epiglottitis

Fig. 12.1   Satellitism. A blood agar plate showing enhancement of the growth of colonies of *Haemophilus influenzae* next to the streak of *Staphylococcus aureus* which supplies V factor.

Antibiotic sensitivity
Sensitive to:
  Ampicillin
  Cotrimoxazole
  Tetracycline
  Erythromycin
  Chloramphenicol (reserve for meningitis and epiglottitis)
  Cefuroxime ⎫
  Cefoxitin   ⎬ second generation cephalosporins
  Cefamandole ⎭

Penicillin: strains are generally resistant to penicillin; in a small proportion, penicillin resistance is due to β lactamase production.

Haemophilus ducreyi
The cause of the sexually transmitted disease *chancroid*, or soft sore.

Laboratory characteristics
  Morphology: slender Gram-negative bacilli, slightly larger than *H. influenzae,* bacteria *en masse* have configuration of 'shoals of fish'.
  Culture: on 20 – 30 per cent rabbit blood agar with CO_2.
  Colonial morphology: small grey glistening colonies surrounded by a zone of haemolysis.
  Growth factors: only X factor is required.

Antibiotic sensitivity: sulphonamide (the drug of choice in treatment).

BRUCELLA

Predominantly infect domestic animals from which infection may be transmitted to man.

Species

Brucella abortus
Brucella melitensis
Brucella suis
  Habitat: chronic infection in domestic animals (see Table 12.3).

Table 12.3 Animal hosts and geographical distribution of *Brucella species*

Strain	Animal host	Geographical distribution
B. melitensis	Goats	Mediterranean countries
B. abortus	Cattle	Worldwide including UK
B. suis	Pigs	Denmark and USA

LABORATORY CHARACTERISTICS

Morphology and staining: slender, pleomorphic, Gram-negative bacilli.

Culture: in enriched medium such as liver infusion broth or agar; CO_2 is required for the growth of *Br. abortus*.

Identification of the different species is done by a variety of biochemical and serological tests and inhibition by dyes: this is illustrated in Table 12.4.

Antigenic structure: the three species share antigens but quantitative differences in the antigens of *Br. melitensis* and *Br. abortus* enable monospecific sera to be prepared to these two species.

PATHOGENICITY

The cause of undulant fever or brucellosis — a chronic, debilitating febrile illness usually without any localising signs: the bacteria persist intracellularly and are therefore difficult to eradicate by antibiotic therapy.

ANTIBIOTIC SENSITIVITY

Tetracycline is the drug of choice.

Table 12.4 Differentiation of Brucella species

Species	Growth requirement for CO_2	Production of H_2S	Growth in presence of		Methyl violet	Agglutination by monospecific antibody to	
			Basic fuchsin	Thionine		Br. abortus	Br. melitensis
B. melitensis	−	−	+	+	+	−	+
B. abortus	+	+	+	−	+	+	−
B. suis	−	+	−	+	−	+	−

BORDETELLA

The only important member of the species is *Bord. pertussis*, the cause of whooping cough.

Bordetella pertussis

Habitat. The human respiratory tract usually associated with acute disease.

LABORATORY CHARACTERISTICS

Morphology and staining: slender, capsulated Gram-negative bacilli.

Culture: special enriched medium and incubation in CO_2 is required: the most widely used media are charcoal blood agar and Bordet-Gengou medium.

Bordet-Gengou medium contains 15 – 20 per cent blood, potato extract, glycerol, agar.

Colonial morphology: colonies like 'split pearls' appear after three or more days of incubation.

Identification is confirmed serologically by slide agglutination.

Antigenic structure: all strains of *Bord. pertussis* possess a predominant surface antigen which causes the agglutination in slide tests: additional surface antigens designated 1 to 6 are also present in varying degrees.

Serotypes: there are three main serotypes based on the presence of the additional numbered surface antigens: type 1, 2; type 1, 2, 3; type 1, 3.

PATHOGENICITY

The cause of whooping cough: a disease mainly seen in pre-school children and especially severe in those under 1 year of age. Affects the lower respiratory tract causing bronchospasm and the characteristic *paroxysmal cough*.

ANTIBIOTIC SENSITIVITY

Erythromycin (also useful prophylactically in susceptible contacts of the disease).

PASTEURELLA

Pasteurella multocida

The main pathogenic member of the genus, also known as *P. septica*.

Habitat. Animals, e.g. dogs.

Morphology: capsulated small, slender Gram-negative bacilli.

Culture: on blood agar (does not grow on MacConkey agar – a differentiating feature from enterobacteria).

Pathogenicity. Causes septic wounds after dog bites and occasionally other forms of trauma.

Antibiotic sensitivity: Gentamicin, tetracyclines — the drugs of choice.

FRANCISELLA

Francisella tularensis
The cause of tularaemia.

Habitat: rodents and small mammals.

Morphology and staining: pleomorphic, small Gram-negative coccobacilli.

Culture: on blood agar enriched with cysteine and glucose.

Pathogenicity. Tularaemia is a plague-like illness contracted by contact with animal hosts or their products: seen in USA and occasionally in Europe.

Antibiotic sensitivity: tetracyclines are the drugs of choice.

ACTINOBACILLUS

Actinobacillus species
Actinobacillus actinomycetem-comitans is sometimes found along with *Actinomyces species* in actinomycosis.

13

Corynebacterium and related bacteria

CORYNEBACTERIUM

Gram-positive bacilli with a characteristic morphology: non-sporing; non-capsulate; non-motile.

Some strains are widely distributed in soil and plants. There are several human and animal species that are important pathogens and commensals.

Corynebacterium diphtheriae
Habitat: the throat and nose of man.

LABORATORY CHARACTERISTICS

Morphology and staining: pleomorphic Gram-positive rods (3 × 0.3 μm) or clubs which divide by 'snapping fission' so that adjacent cells lie at different angles to each other forming V, L and W shapes — groups of such cells produce a *Chinese-character* arrangement; adjacent cells may also lie parallel to one another in *palisades*.

Cells stain irregularly due to intracellular deposition of polymerised phosphate forming the metachromatic or volutin granules that are characteristic of, but not exclusive to, *C. diphtheriae*. The granules, usually two or three per cell, show up with special stains — bluish black by Albert's method, deep blue with Neisser's methylene blue.
Culture: an aerobe and facultative anaerobe; optimum temperature 37°C. Does not grow well on ordinary agar and media containing blood or serum are required. Selective media are necessary for isolation from clinical specimens.

Selective media:
1. *Loeffler's serum medium: C. diphtheriae* grows rapidly, faster than other upper respiratory tract bacteria present in clinical material: the morphology develops particularly well and smears made as soon as 8 hours after inoculation may show a typical appearance (Table 13.1).

2. *Blood Tellurite agar* (e.g. Hoyle's or McLeod's medium): after 48 h incubation, corynebacteria produce characteristic grey-black colonies due to their ability to reduce potassium tellurite to tellurium.

Three colonial types of *C. diphtheriae* are recognised, *gravis*, *intermedius* and *mitis*; they were so named from the severity of the clinical disease for which they were responsible, but the correlation with colonial morphology is not good (Table 13.1). Other corynebacteria also grow on tellurite media to form colonies that can be confused with *C. diphtheriae*.

Table 13.1   Characteristics of *Corynebacterium diphtheriae* strains

Corynebacterium diphtheriae	Appearance in film from Loeffler's medium	Colonial type on tellurite medium
Gravis strains	Club-shaped, few metachromatic granules	Flat, grey with raised centre and irregular edge. Radial striations develop to form a 'daisy-head'
Intermedius strains	Short irregularly staining rods without metachromatic granules but in Chinese character arrangement	Small, smooth colonies of uniform size; grey-black with paler periphery
Mitis strains	Classic morphology with numerous granules and typical arrangement	Medium-sized, circular convex, glistening and black

Identify: by biochemical tests and demonstration of toxin production. Other corynebacteria may mimic *C. diphtheriae* in films and on culture. Some isolates of *C. diphtheriae*, especially mitis strains, are not toxigenic and therefore non-virulent.

Biochemical reactions: acid production from a range of carbohydrates and other biochemical tests are used to differentiate *C. diphtheriae* from other corynebacteria. Gravis (but not intermedius or mitis) strains ferment starch and glycogen.

Typing. Serotyping by agglutination tests, phage typing and bacteriocin typing have all been used to subdivide strains of *C. diphtheriae* for epidemiological studies.

Toxin production: is responsible for virulence; can be demonstrated by guinea pig inoculation or by a gel precipitation test.

1. *Guinea pig inoculation:* inject suspension of the isolated strain of *C. diphtheriae* into two guinea pigs, one protected with diphtheria antitoxin.

Result	Unprotected animal	Antitoxin protected animal
Toxigenic strain	Death in 2-3 days	Survival
Non-toxigenic strain	Survival	Survival

2. *Gel-precipitation (Elek) test.* A filter paper strip previously immersed in diphtheria antitoxin is incorporated into serum agar before it has set; the strain of *C. diphtheriae* under investigation is then streaked onto the agar at right angles to the filter paper strip. Incubate at 37°C.

Observe after 24 h and 48 h for precipitation indicating toxin–antitoxin interaction (Fig. 13.1).

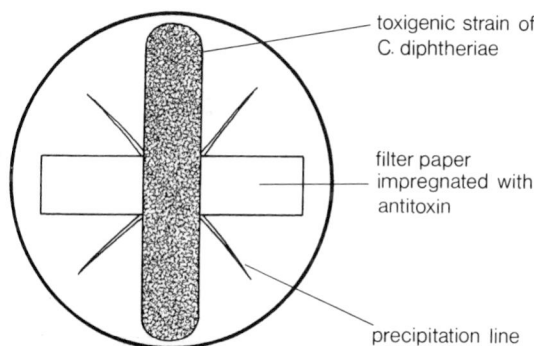

toxigenic strain of
C. diphtheriae

filter paper
impregnated with
antitoxin

precipitation line

Fig. 13.1 Elek test for demonstration of toxin production by *Corynebacterium diphtheriae*. The toxin combines with antitoxin to produce antigen–antibody complexes which form visible lines of precipitation in the agar.

Diphtheria toxin. The exotoxin of *C. diphtheriae* is produced only by strains carrying a bacteriophage: its formation *in vitro* is stimulated in culture media with low iron content. The toxin interferes with protein biosynthesis in mammalian cells by splitting the molecule of NAD (nicotinic adenine dinucleotide) — an essential cofactor for the transferase involved in peptide bond formation by ribosomes. It acts locally on the mucous membranes of the respiratory tract to produce a grey adherent pseudomembrane consisting of fibrin, bacteria, epithelial and phagocytic cells: after adsorption into the bloodstream, it acts systemically on the cells of the myocardium, the nervous system (only motor nerves are affected) and the adrenals.

The toxin can be rendered non-toxic but still antigenic by treatment with formaldehyde: the *toxoid* so formed is used in prophylactic immunisation.

Schick test: is a skin test to demonstrate circulating diphtheria antitoxin — the result either of previous immunisation or infection (clinical or subclinical).

Method: intradermal injection of toxin into the anterior aspect of one forearm and heat-inactivated toxin into the other.

Object: to detect susceptibility to diphtheria toxin and hypersensitivity to the toxin or other more heat-stable proteins produced by *C. diphtheriae*. In children over 10 years of age and in adults, a preliminary Schick test should be carried out before immunisation to detect pseudo-reactors who may experience severe allergic responses to the vaccine.

Observe: for erythema at the injection site: reactions due to the toxin are slower to develop and longer lasting than those resulting from hypersensitivity. The arm should be examined at 1–2 days (e.g. 36 hours) and again at 5–7 days (e.g. 120 hours). The reactions are summarized in Table 13.2.

Table 13.2   The Schick test

Result	Test arm (toxin) 36h	120h	Control arm (inactivated toxin) 36h	120h	Interpretation	Diphtheria immunisation
Negative	—	—	—	—	Immune, not hypersensitive	Not required
Positive	±	+	—	—	Non-immune, not hypersensitive	Required
Negative and pseudo	+	—	+	—	Immune, hypersensitive	Not required
Positive* and pseudo 'combined'	+	+	+	—	Non-immune, hypersensitive	Contra-indicated

*Combined reactions are rare.

PATHOGENICITY

C. diphtheriae is the cause of diphtheria: usually the mucous membranes of the upper respiratory tract are affected but sometimes, especially in tropical countries, skin lesions are produced: the serious systemic manifestations follow absorption of the exotoxin.

ANTIBIOTIC SENSITIVITY

C. diphtheriae is sensitive to penicillin, erythromycin, the lincomycins and other antibiotics.

OTHER CORYNEBACTERIA

Corynebacterium ulcerans
May be responsible in man for diphtheria-like throat lesions but with little evidence of toxaemia.
   Biochemical reactions distinguish it from *C. diphtheriae*.
   Toxins: two are produced — one immunologically identical to the toxin of *C. diphtheriae*, the other related to the toxin of *C. ovis*.

Animal pathogens
A number of corynebacteria (*C. ovis, C. murium, C. pyogenes, C. equi, C. renale*) are important animal pathogens. The first two are morphologically similar to *C. diphtheriae*.

Human commensals
There are many so-called *diphtheroid bacilli*, the most often isolated being *C. hofmannii* in the throat and *C. xerosis* in the conjunctiva.
   Habitat: normally present in the skin (especially within sebaceous ducts) and mucous membranes.
   Morphology and staining: less pleomorphic and more strongly Gram-positive than *C. diphtheriae*; metachromatic granules are few or absent. Tend to be arranged in pallisades with less pronounced Chinese lettering.
   Culture: grow well on ordinary agar.
   Pathogenicity: occasional opportunist pathogens, e.g. causing endocarditis on prosthetic valves, infection in implanted artificial joints.

PROPIONIBACTERIUM

Gram-positive, non-sporing, anaerobic bacilli formerly classified as anaerobic corynebacteria. A differential feature from other similar anaerobes is that they produce propionic acid as a breakdown product of carbohydrate fermentation; this can be detected by gas-liquid chromatography.

Two main species are recognised, *Propionibacterium acnes* and *P. granulosum*.

Habitat: the human skin.

Pathogenicity: apparent association with the skin disease *acne vulgaris*.

ERYSIPELOTHRIX

Erysipelothrix rhusiopathiae
A slender Gram-positive non-motile bacillus, sometimes filamentous: related to the corynebacteria.

Habitat: healthy pigs but widely distributed in other animals and birds: found on the skin and scales of fish: *causes* swine erysipelas.

Pathogenicity: responsible in man for erysipeloid — a rare skin infection.

LISTERIA

Listeria monocytogenes
Morphologically similar to erysipelothrix but flagellated; feebly motile at 37°C but exhibits active tumbling motility at 25°C in young broth cultures.

Habitat: wild and domestic animals.

Culture: aerobic and facultatively anaerobic; optimal temperature 37°C but will survive and grow at 5°C.

Colonies on blood agar surrounded by a narrow zone of complete (β) haemolysis.

Pathogenicity: causes listeriosis in man.

14

Mycobacterium

Mycobacteria are widely distributed in nature: referred to as acid-fast because after mordanting in stain they resist decolorisation with strong acids.

The main medically important mycobacteria are shown in Table 14.1 together with some of their properties.

Table 14.1 Medically important mycobacteria

Species	Habitat and source	Disease	Cultural characteristics on Lowenstein-Jensen medium
Myco. tuberculosis	Infected humans	Tuberculosis	Rough, dry yellow colonies
Myco. bovis	Infected cattle	Tuberculosis	White, smooth colonies (inhibited by glycerol)
Myco. leprae	Infected humans	Leprosy	No growth
Atypical mycobacteria	Mainly soil, water, sometimes birds, fish	Pulmonary infection, cervical adenitis, skin ulcers	Often pigmented colonies: some species grow rapidly; others have unusual temperature requirements

Mycobacterium tuberculosis

LABORATORY CHARACTERISTICS

Morphology: slender, beaded bacilli: aerobic, non-sporing.

Staining: does not stain by Gram's method. Ziehl-Neelsen stain must be used: smears are treated with concentrated carbol fuchsin, mordanted by heating and then decolorised with 20 per cent sulphuric acid and alcohol; mycobacteria retain a bright red colour after this treatment and show up clearly against the background of counterstain (usually malachite green or methylene blue).

Culture: aerobe: does not grow on ordinary media: grows on Lowenstein-Jensen medium (contains egg, asparagine, glycerol and — to inhibit contaminants — malachite green) usually after 2–3 weeks' incubation at 37°C; cultures should be kept for 6–8 weeks before being discarded.

PATHOGENICITY

The cause of tuberculosis — a slowly progressive, chronic infection usually of the lungs but many other organs and tissues may become affected: the main source of infection is respiratory secretions from a patient with 'open' disease.

Pathogenic for guinea pigs, with production of typical progressive disease and death in 6–8 weeks; rabbits are much less susceptible.

Mycobacterium bovis
Main source of infection is infected cattle: man becomes infected by ingestion of milk containing *Mycobacterium bovis*.

Pathogenicity: similar to tuberculosis due to human tubercle bacilli: particularly liable to infect children with lesions in cervical lymph nodes, bones and joints, kidneys.

Pathogenic for both guinea pigs and rabbits with development of generalised disease.

Rare nowadays due to eradication of disease in cattle.

ANTIBIOTIC SENSITIVITY

Both *Myco. tuberculosis* and *Myco. bovis* are sensitive to a wide range of drugs (listed below); because resistant variants arise readily, therapy should always be a combination of three drugs.

Isoniazid
Rifampicin
Ethambutol
Streptomycin
Pyrazinamide
Ethionamide
Thioecetazone
Cycloserine

Mycobacterium leprae
The cause of leprosy — still a scourge in many parts of the Third World today.

Habitat: the organism is found only in cases of human infection.

Cultivation: in vivo by inoculation of the footpads of mice: mice develop slow-growing granulomas at the site of injection. Does not grow *in vitro.*

Antibiotic sensitivity

Sensitive to:
  Dapsone (a sulphone)
  Rifampicin
  Clofazimine

Atypical mycobacteria

A group of miscellaneous mycobacteria of low pathogenicity for man.

Culture: generally grow on Lowenstein–Jensen medium sometimes at lower (25°C) or higher (45°C) temperatures than normal. Several species produce pigmented growth and this may be increased by exposure to light (photochromogens).

Classification: controversial; the original Runyon groups I-IV now not often used.

Species. The principal species found with some of their features are shown in Table 14.2.

Table 14.2 Atypical mycobacteria

Species	Disease	Normal habitat
Myco. kansasii	Pulmonary Cervical adenitis	Water
Myco. avium/ intracellularle	Pulmonary Cervical adenitis	Birds, soil
Myco. xenopi	Pulmonary	Water
Myco. marinum	Skin ulcers	Water, tropical fish tanks
Myco. fortuitum	Cervical adenitis	Soil
Myco. scrofulaceum	Cervical adenitis	Soil
Myco. ulcerans	Tropical skin ulcers	Soil

Pulmonary infection. In many cases, atypical mycobacteria are simply 'passengers' which accompany tuberculosis, and treatment — even when the atypical mycobacteria are resistant — usually results in cure.

Antibiotic sensitivity. Variable, often resistant to several of the standard antituberculous drugs.

15

Actinomyces and nocardia

Actinomyces and nocardia are morphologically similar Gram-positive branching rods and filaments. Actinomyces are microaerophilic or anaerobic on primary isolation although some species grow in air after a few subcultures; nocardia are aerobic organisms.

ACTINOMYCES

Most actinomyces are soil organisms but some — and these are the potentially pathogenic species — are commensals of the mouth of man and animals (Table 15.1).

Table 15.1   Some Actinomyces species

Species	Host/habitat	Disease association
Actino. israelii	Oropharynx and gut of man	Human actinomycosis
Actino. naeslundii *Actino. viscosus* *Actino. odontolyticus* }	Oropharynx of man	Dental plaque and caries
Actino. bovis	Oropharynx of cattle	Lumpy jaw in cattle

Species are identified by colonial appearances (some are pigmented), ability to grow aerobically and biochemical tests.

Actinomyces israelii

LABORATORY CHARACTERISTICS

Morphology and staining. Gram-positive bacteria which grow in filaments that readily break up into rods and may show branching. Non-motile, non-sporing, not acid-fast. In tissue, colonies develop to form diagnostic yellowish 'sulphur granules' which are visible to the naked eye and which are found in pus discharged through draining sinuses.

Culture:
Solid media: on blood or serum glucose agar incubated anaerobically at 37°C for 7 days or more; growth is enhanced by 5 per cent carbon dioxide.
Observe: for small, cream, adherent, nodular colonies.
'Shake' cultures in semi solid glucose agar kept at 37°C for 5–10 days.
Observe: for maximal growth in a turbid band 10–15 mm below the surface where conditions are microaerophilic.

Isolation of this exacting microorganism from clinical material is difficult especially as the pus often contains other faster-growing bacteria. Presumptive diagnosis is made by the demonstration of typical Gram-positive branching filaments in a sulphur granule. Whenever possible, a washed, crushed sulphur granule should be cultured in preference to pus.

PATHOGENICITY

Actinomycosis is endogenous in origin and results in a chronic granulomatous infection with abscess formation: profuse pus discharges by draining through sinuses. Infection probably starts after local trauma, e.g. the extraction of carious teeth, appendicectomy. The typical sites of the disease are: cervico-facial — 65 per cent of cases: abdominal (usually ileo-caecal) — 20 per cent of cases; and, rarely, thoracic, affecting the lung.

ANTIBIOTIC SENSITIVITY

Sensitive to penicillin, lincomycins, tetracyclines, erythromycin.

NOCARDIA

Habitat: majority of species are soil saprophytes; a few are pathogenic to man.
Morphology and staining: similar to actinomyces but some species are weakly acid-fast.
Culture: slow-growing, aerobic organisms which require 5–14 days incubation on a nutrient agar.
Observe: wrinkled, rosette to star-shaped colonies, initially white, then yellow and finally pink or red.

Antibiotic sensitivity: sensitive to sulphonamides and cotrimoxazole: resistant to practically all antibiotics. Treatment may have to be continued for six months.

Pathogenicity: cause chronic granulomatous suppurative infections.

Nocardia asteroides
Affects lungs, sometimes with secondary brain abscess.

Nocardia madurae and Nocardia brasiliensis
Affect subcutaneous tissues: 'madura foot' or mycetoma is a tropical form of nocardiosis which affects the foot and produces chronic discharging sinuses.

Opportunistic infection: pulmonary nocardiosis has been described in renal transplant and other immunocompromised patients.

16

Neisseria

The neisseriae are Gram-negative diplococci: the two pathogenic species *Neisseria gonorrhoeae* (the gonococcus) and *Neisseria meningitidis* (the meningococcus) have exacting growth requirements; the commensal neisseriae are easy to culture.

Neisseria gonorrhoeae
Habitat: an obligate parasite of the human urogenital tract.

LABORATORY CHARACTERISTICS

Morphology and staining. Gram-negative cocci 0.6–1.0 μm in diameter, occur singly but characteristically in pairs with adjacent sides flattened or concave ('bean-shaped') and long axes parallel. In purulent clinical material many of the diplococci are intracellular within a relatively small number of the polymorphs, the remainder extracellular in the exudate. Some pleomorphism and variation in the intensity of staining is common.

Culture: requires an enriched medium (usually lysed blood agar or chocolate agar) and incubation in a moist aerobic atmosphere containing 5–10 per cent CO_2. Optimal temperature 35–36°C.

Selective media: chocolate agar can be made selective by the addition of antibiotics which inhibit other bacteria but not *N. gonorrhoeae*, e.g. Thayer–Martin medium which contains vancomycin, colistin, nystatin and trimethoprim. Used for the isolation of *N. gonorrhoeae* from clinical material especially when a wide range of other bacteria will be present, e.g. specimens from vagina and rectum.

Observe: small grey glistening colonies after 24 h incubation becoming larger, opaque and somewhat irregular at 48 h.

Recognition of colonies in mixed cultures: N. gonorrhoeae (and *all* other neisseriae) give a positive oxidase test.

Method: pick suspect colonies onto filter paper moistened with 1 per cent solution of tetramethyl-*p*-phenylenediamine (the 'oxidase reagent'). Oxidase-positive colonies rapidly turn the paper a dark purple colour. (This method is preferable to flooding the plate with oxidase reagent and observing the development of a dark-purple colour with oxidase-positive colonies which then have to be subcultured without delay because the solution is quickly lethal.)

Identification: by carbohydrate utilization tests; *N. gonorrhoeae* produces acid from glucose only and no other sugar substrate.

Antigenic structure. Complex, cross-reactivity with other neisseriae: antigens in pili, cell wall lipopolysaccharide and outer membrane proteins; the latter may be strain-specific. Varies greatly with cultural conditions.

Typing: no definitive method.

PATHOGENICITY

The cause of the venereal disease gonorrhoea, a pyogenic infection of the urethra and, in females, of the uterine cervix.

Viability: dies rapidly outside the human host but may remain viable in pus for some time.

ANTIBIOTIC SENSITIVITY

Sensitive to penicillin, ampicillin, tetracycline, macrolides, spectinomycin, cotrimoxazole, chloramphenicol, cefuroxime and other drugs. Low-level penicillin resistance is present in 10–20 per cent of strains isolated in Britain but this does not preclude treatment with the drug. Since 1976, however, highly resistant β-lactamase producing strains have been encountered abroad and some have caused disease in the United Kingdom.

Neisseria meningitidis

Habitat: the human nasopharynx: present in 5–10 per cent of normal people.

LABORATORY CHARACTERISTICS

Morphology and staining: as for *N. gonorrhoeae.* Films of the CSF in meningococcal meningitis have the same appearance as genital tract exudate in gonorrhoea.

Culture: requirements very similar to those described for *N. gonorrhoeae* although somewhat less exacting. On culture, the colonies are slightly larger. The use of an antibiotic-containing selective medium facilitates isolation of *N. meningitidis* from the normal mixed pharyngeal flora. The oxidase test is an aid to the recognition of suspect colonies but note that the commensal neisseriae will also give a positive reaction.

In meningitis the organism is present in pure culture.

Identification: by carbohydrate utilization; *N. meningitidis* produces acid from maltose and glucose.

Antigenic structure. Although antigens are shared with other neisseriae recognition of polysaccharide antigens has allowed differentiation into serological groups. There are three main groups (A, B, C) and five subsidiary groups (X, Y, Z, 29E and W135): not all isolates are groupable.

Typing: serological typing into the above groups by slide agglutination with specific antisera.

PATHOGENICITY

Cause of meningococcal meningitis: in those susceptible, spread from the nasopharynx results in a septicaemia which is usually followed by rapid involvement of the meninges. However, chronic meningococcal septicaemia without meningitis is a recognised, although rare, clinical entity.

Viability: dies quickly at room temperature outside the human host.

ANTIBIOTIC SENSITIVITY

Sulphonamide sensitivity was formerly the rule but resistance is now common: in Britain, at least 10 per cent of strains are fully resistant and a further 50 per cent partially resistant.

Sensitive to penicillin, ampicillin, chloramphenicol, tetracyclines, macrolides and other drugs.

Commensal neisseriae

Habitat: regularly present in the mucous membranes of the mouth, nose and pharynx; less frequently in the genital tract.

Members of this heterogeneous collection of oxidase-positive Gram-negative diplococci can be differentiated from the two pathogenic species because of their ability to grow: (i) on ordinary agar not supplemented with blood or serum; (ii) at 22°C (room-temperature); (iii) on primary isolation in the absence of CO_2.

Classification can be simplified by their division into two groups:

1. *Neisseria pharyngis* group (includes *N. flava, N. perflava, N. sicca* and *N. subflava*).

   Colonies: often pigmented (yellow to green) and rough.

   Biochemical activity: characteristically produce acid from glucose and maltose, sometimes from sucrose but not from lactose.

2. *Neisseria catarrhalis* group recently placed in a separate genus as *Branhamella catarrhalis*).
Colonies: colourless but of n opaque and sometimes rough.
Biochemical activity: fail to oroduce acid from carbohydrates

Pathogenicity
With rare exceptions they are ιot responsible for disease.

RELATED BACTERIA

Members of the genus *Mo xella* and the ge us *Acinetobacter* (considered in Chapter 10) are ·acilli bu may exh it coccobacillary morphology. For a varie⁺v of axonon c reasons they are closely related to e⌐ch other and to *N sse⁻ia* ə d *Branho ella*. Superficial resemblance nay lead to confu ın and difficultie in identification.

17

Bacillus

Members of the genus *Bacillus* are aerobic, sporing, Gram-positive, chaining bacilli. *Bacillus species* are ubiquitous saprophytes but one, *Bacillus anthracis*, is an important pathogen responsible for anthrax in animals and man.

Bacillus anthracis

Habitat: infected animals but spores are found in soil and pasture contaminated with vegetative cells from dead and dying animals.

LABORATORY CHARACTERISTICS

Morphology: large (4–8 × 1.5 μm), non-motile, rectangular bacilli usually arranged in chains; spores are oval and central — not formed in tissues but developed after the organism is shed or grown on artificial media: capsulated in the animal body and on laboratory culture under certain conditions: the capsule consists of a polypeptide of D-glutamic acid.

Staining: Gram-positive; spores can be stained using modified Ziehl-Neelsen method.

McFadyean's reaction: used to demonstrate *B. anthracis* in the blood of animals in a heat-fixed film stained with polychrome methylene blue.

Observe for blue bacilli surrounded by purplish-red amorphous material due to disintegrated capsules and indicating a positive reaction: diagnostic of *B. anthracis*.

Culture: aerobe and facultative anaerobe; grows readily on ordinary media over a wide temperature range (optimum 35°C): best temperature for sporulation is lower 25–30°C.

Colonies are large, dense, grey-white, matt and irregular: they are composed of parallel chains of cells and this gives the margin of the colony the so-called 'medusa head' or 'curled hair lock' appearance.

Blood agar: there is only slight haemolysis round the colony — a differential feature because other *Bacillus species* are markedly haemolytic.

113

Broth cultures develop a thick pellicle.

Gelatin stab cultures show growth along the track of the wire with lateral spikes longest near the surface — the 'inverted fir tree'; liquefaction is late, starting at the surface.

Antigenic structure: the antigenic components described include:

(i) a complex group of toxins
(ii) the capsular polypeptide

PATHOGENICITY

A wide range of animal hosts are susceptible; infection is characteristically septicaemic with splenic enlargement. Man is infected from animals or animal products.

Viability: vegetative cells are readily destroyed by heat but spores demonstrate a variable but often high level of heat resistance — in the dry state, up to 150°C for 1 h. Spores can remain viable for years in contaminated soil.

ANTIBIOTIC SENSITIVITY

B. anthracis is susceptible to many antibiotics: penicillin is the drug of choice.

OTHER BACILLUS SPECIES ('ANTHRACOID BACILLI')

Habitat: saprophytes in soil, water, dust and air and on vegetation.

Many species are recognised: some (e.g. *B. megatherium, B. cereus*) are large-celled like *B. anthracis*, others (e.g. *B. subtilis*) are small-celled and shorter and thinner with rounded ends. They differ from *B. anthracis* in being motile, non-capsulated and by the distinct zone of haemolysis round colonies on blood agar: furthermore, they fail to produce a fatal septicaemia in laboratory animals.

The spores of certain anthracoid bacilli are used as a test of the efficiency of sterilisation by steam under pressure (the autoclave), by ethylene oxide and by ionizing radiation.

Bacillus cereus is a cause of food-poisoning (e.g. when contaminating rice).

18

Clostridium

Clostridia are anaerobic, sporing, Gram-positive bacilli. Most species are soil saprophytes but a few are pathogens. The most important are listed with some of their principal properties in Table 18.1.

Table 18.1 Pathogenic properties of the main medically-important species of Clostridia

Species	Disease
Cl. welchii	Gas gangrene, food-poisoning
Cl. oedematiens	Gas gangrene
Cl. septicum	Gas gangrene
Cl. histolyticum	Secondary role in gas gangrene
Cl. fallax	Secondary role in gas gangrene
Cl. bifermentans	Secondary role in gas gangrene
Cl. sordellii	Secondary role in gas gangrene
Cl. difficile	Antibiotic-associated colitis
Cl. sporogenes	Doubtful pathogenicity
*Cl. tetani**	Tetanus
Cl. botulinum	Botulism

*forms round, terminal spores; the other species have oval subterminal spores.

Habitat: human and animal intestine; soil, water, decaying animal and plant matter.

LABORATORY CHARACTERISTICS

Morphology and staining: large (3–8 × 0.5 μm) rods, sometimes pleomorphic, filamentous forms common: Gram-positive but may stain irregularly or be Gram-negative in older cultures.

Spores: all species form endospores which are usually 'bulging', i.e. wider than the bacterial body; sometimes useful in identification, e.g. *Cl. tetani*. *Cl. welchii* (the most common human pathogen) forms spores with difficulty.

Motile: with peritrichous flagella (*Cl. welchii* is non-motile).

Capsule: Cl. welchii has a capsule but most are non-capsulated.

Culture:

1. Blood agar anaerobically: in mixed culture, addition of an aminoglycoside makes an excellent selective medium for clostridia.
2. Robertson's meat medium.

Anaerobic requirement: variable. *Cl. tetani* and *Cl. oedematiens* are exacting anaerobes; *Cl. welchii* and *Cl. histolyticum* can grow in the presence of limited amounts of oxygen.

Colonial morphology: the main human pathogenic species show three types of colonial morphology.

a. *Cl. welchii:* round opaque colonies surrounded by a zone of β haemolysis.
b. *Cl. tetani: Cl. septicum:*
   Irregular, translucent colonies with spreading edges; marked tendency to swarm especially with *Cl. tetani.*
c. *Cl. oedematiens: Cl. sporogenes: Cl. botulinum:*
   Translucent colonies with irregular margins and filamentous outgrowths.

Biochemical activity

Saccharolytic: many species ferment sugars; this produces reddening of the meat particles in Robertson's meat medium with a rancid smell.

Proteolytic: production of enzymes that digest proteins is a common property of many clostridia: this causes blackening and digestion of the meat particles in Robertson's meat medium with a foul smell; gelatin and coagulated serum are liquefied.

Toxins: medically-important clostridial species produce several toxins: the exotoxins of *Cl. tetani* and *Cl. botulinum* are amongst the most toxic substances known: clostridial toxins are often haemolytic and lethal for various species of laboratory animal.

Identification: difficult: the identification of the three most important pathogenic clostridia — namely *Cl. welchii, Cl. tetani* and *Cl. botulinum* — is described below.

ANTIBIOTIC SENSITIVITY

Sensitive to penicillin, metronidazole, and other antibiotics such as clindamycin, tetracycline, erythromycin: resistant to aminoglycosides.

Clostridium welchii (Clostridium perfringens)

Laboratory characteristics

Morphology: a short stubby bacillus in which spores are hardly ever seen.

Culture: most strains grow well on blood agar anaerobically producing β-haemolytic colonies but some strains are non-haemolytic.

Biochemical activity: mainly saccharolytic: in tube cultures of litmus milk a characteristic 'stormy clot' is formed due to the production of acid and large amounts of gas.

Typing: Cl. welchii can be divided into five types — A, B, C, D and E — on the basis of the toxins formed; all five types produce α toxin. Type A is the human pathogen.

Toxins:

Alpha (α) toxin: an enzyme, phospholipase C, which causes cell lysis due to lecithinase action on the lecithin in mammalian cell membranes: formed by all strains of *Cl. welchii* but in greatest quantity by type A.

Other toxins: include collagenase, proteinase, hyaluronidase, deoxyribonuclease. Several have haemolytic activity and are described as 'necrotising' or 'lethal' from their effects on laboratory animals.

Identification:

Nagler reaction: identifies *Cl. welchii* by neutralisation of α toxin by specific antitoxin: colonies are streaked on agar plates containing egg yolk (egg yolk contains lecithin) half of the plate having been spread with antitoxin: a dense opacity is produced by the growth of *Cl. welchii* on the untreated half of the plate but there is no opacity on the area with antitoxin (Fig. 18.1).

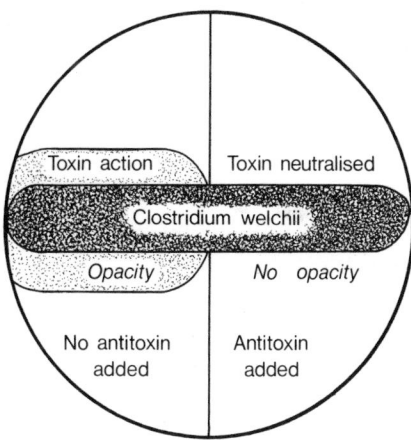

Fig. 18.1   Nagler reaction. The alpha toxin (a lecithinase) of *Clostridium welchii* has produced opacity due to degradation of lecithin in the medium on the left. This action has been neutralised by the antitoxin on the right of the plate.

Pathogenicity

Food-poisoning: when ingested in large numbers, *Cl. welchii* produces an enterotoxin in the gut causing diarrhoea and other symptoms of food poisoning.

Gas gangrene: wounds associated with necrosis of muscle may become infected with *Cl. welchii* and other clostridia causing a severe and life-threatening spreading infection of the muscles.

Clostridium tetani

Morphology: a longer, thinner bacillus with round terminal spores giving characteristic 'drum-stick' appearance (Fig. 18.2).

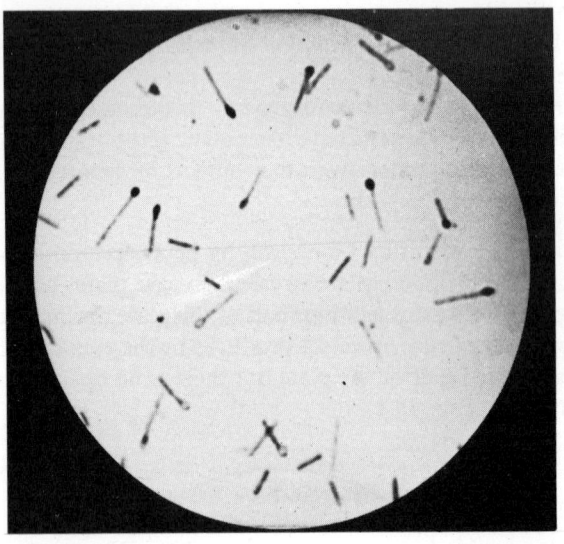

Fig. 18.2   Tetanus bacilli. Showing characteristic drumstick morphology due to round terminal spores (magnified × 2000).

Toxin: a protein and exceedingly potent; two components.

(i) tetanospasmin: neurotoxic
(ii) tetanolysin: haemolytic

Identification: Toxin neutralisation *in vivo:* a culture filtrate of the test organism is injected into mice, some of which have been protected by previous inoculation of tetanus antitoxin. In a positive result the unprotected animals die in a few days with typical tetanic spasms; protected animals survive.

Typing: there are 10 serological types of *Cl. tetani*; all produce the same toxin.

Pathogenicity
The cause of *tetanus*, a classical toxin-mediated disease in which *Cl. tetani* in a wound elaborates the powerful neurotoxin which spreads and acts on the central nervous system causing severe muscle spasms.

Clostridium botulinum

Habitat: soil — virgin and cultivated, sand and sea-water sludge, rarely animal gut.

Toxin: protein and even more potent than that of *Cl. tetani*, the toxin of *Cl. botulinum* in the most active known poison; it acts by preventing release of acetylcholine at motor nerve endings in the parasympathetic system; destroyed at 80°C for 30 min.

Typing: six serotypes — A, B, C, D, E and F each with a serologically distinct toxin.

Human botulism is due to types A, B and E.

Identification: by testing in mice for neutralisation of toxin in a culture, patient's serum or food sample by specific antitoxin. Some of the mice are injected with antitoxin to A, B and E toxins; mice inoculated with the appropriate antitoxin survive, the others become paralysed and die.

Pathogenicity: produces a rare form of 'food-poisoning' known as botulism in which the symptoms are neurological rather than intestinal; due to ingestion of pre-formed toxin in food contaminated with the organism.

Clostridium difficile

Recently discovered as the main cause of antibiotic-associated pseudomembranous colitis.

Habitat: apparently rare in faeces of healthy adults but regularly found with its toxin in the faeces of patients suffering from acute pseudomembranous colitis resulting from administration of antibiotics (most commonly clindamycin).

Identification:

1. *Isolation:* from faeces using selective media, e.g. cefoxitin-cycloserine agar: anaerobic incubation.
2. *Demonstration of toxin* in faeces by inoculation of cell cultures (e.g. HeLa or human embryo lung cells) with faecal extract: cultures containing antitoxin to the toxin of *Cl. sordellii* — a serologically similar species — are included.

   Observe: for cytotoxicity which is neutralised in the cultures containing *Cl. sordellii* antitoxin.

19

Bacteroides and other non-sporing anaerobes

The non-sporing anaerobes are listed in Table 19.1.

Table 19.1 Non-sporing anaerobes

	Gram-positive	Gram-negative
Bacilli	Bifidobacterium	Bacteroides
	Propionibacterium*	Fusobacterium
	Eubacterium	Leptotrichia
Cocci	Peptococcus	Veillonella
	Peptostreptococcus	

*Described in Chapter 13.

BACTEROIDES AND FUSOBACTERIA

The classification of these organisms is controversial: Table 19.2 is a simplified list of some of the main species.

Habitat:

Colon: bacteroides are present in enormous numbers in the faeces (10^{10} per gram or more). Almost all (about 80 per cent) belong to the fragilis group but only a minority, 1 in 10, are *Bacteroides fragilis*: another 10 per cent belong to the asaccharolytic group.

Table 19.2 *Bacteroides* and *Fusobacterium* species

Bacteroides			Fusobacterium
Group			
Fragilis	Melaninogenicus/ oralis	Asaccharolytic	
B. fragilis	*B. melaninogenicus*	*B. asaccharolyticus*	*F. nucleatum*
B. vulgatus	*B. oralis*		*F. necrophorum*
B. distasonis			*F. necrogenes*
B. ovatus			(*F. fusiforme**)
B. thetaiotaomicron			

*Now re-classified as *Leptotrichia buccalis*.

Female genital tract: bacteroides are common in the cervix and vaginal fornices: about 80 per cent belong to the melaninogenicus/oralis group; 15 per cent belong to the asaccharolytic group; *B. fragilis* is uncommon in this site.

Mouth: always present in large numbers in the normal mouth. About 70 per cent of strains belong to the melaninogenicus/oralis group, the commonest species being *B. oralis:* another 20 per cent of strains are fusobacteria.

LABORATORY CHARACTERISTICS

Morphology and staining: Gram-negative, non-motile, non-sporing bacilli: bacteroides are usually small, ovoid or short bacilli. Fusobacteria tend to be long and spindle-shaped: pleomorphism is common.

Culture: strict anaerobes: require media enriched with blood or haemin: the growth of many strains is improved by the addition of menadione (vitamin K3).

1. *Fluid media:* Robertson's cooked meat broth, preferably enriched, e.g. with yeast extract: the medium should be boiled and promptly cooled before use to remove dissolved oxygen and so improve anaerobiasis.
2. *Selective media:* incorporation of antibiotics to which bacteroides are resistant aids isolation from mixed cultures, e.g. blood agar containing an aminoglycoside (neomycin, kanamycin, gentamicin) or an aminoglycoside and vancomycin.

   Incubate: anaerobically with 10 per cent carbon dioxide (which enhances growth) for a minimum of 48 h or longer.

Observe:

1. *Fragilis group:* light grey, opaque or translucent colonies usually 1–2 mm in diameter after 48 h incubation.
2. *Melaninogenicus-oralis group:* slower growing than the fragilis group; *B. melaninogenicus* characteristically produces black or brown pigmented colonies with haemolysis on blood agar; other species in this group are non-pigmented.
3. *Asaccharolytic group:* slower growing than the fragilis group; *B. asaccharolyticus* produces black or brown colonies on blood agar.
4. *Fusobacteria:* some species produce dull granular colonies which may be rhizoid.

Identification: most anaerobes can be identified to generic level by examination of a Gram-stained smear and by gas chromatographical analysis of fatty acid end-products of glucose metabolism. *Bacteroides* and *fusobacterium species* can be identified by a series of tests including colonial morphology, biochemical tests, growth inhibition by bile salts and antibiotic resistance. Species differentiation is time-consuming and costly: most laboratories simply report isolation of bacteroides.

PATHOGENICITY

The most common isolate from clinical specimens is *B. fragilis* and thus this species seems to have a special pathogenic potential.

Bacteroides are important in abdominal and gynaecological (including puerperal) sepsis: they are usually found along with other organisms, notably coliform organisms. Fusobacteria play an important role in dental sepsis in association with melaninogenicus/oralis strains and borrelia.

It appears that these combinations of bacteria potentiate the ability of each other to cause infection — *pathogenic synergy*.

ANTIBIOTIC SENSITIVITY

Like all anaerobic organisms bacteroides and fusobacteria are sensitive to metronidazole: many are also sensitive to the lincomycins, chloramphenicol and cefoxitin. Members of the fragilis group are penicillin-resistant, but many strains of the other groups are penicillin-sensitive. There is uniform resistance to the aminoglycosides.

BIFIDOBACTERIUM

Bifidobacterium is a genus of diverse, *non-pathogenic* bacteria. *Bifidobacterium bifidum* is the types species. Previously they were classified as 'anaerobic lactobacilli'.

Habitat: dominant members of the colonic flora of infants (classically the breast-fed) and common also in the adult gut; normally present in the human vagina.

Laboratory characteristics
Morphology and staining: pleomorphic Gram-positive bacilli characterised by club-shaped rods and by branching forms which are often Y-shaped; non-sporing, non-motile.
Culture: strict anaerobes.

EUBACTERIUM

Habitat: commensal in the human intestine and also in the mouth: sometimes isolated in mixed culture from clinical material.

Morphology and staining: Gram-positive bacilli, often small but size varies with species: non-sporing, non-motile.

ANAEROBIC COCCI

The classification of this heterogeneous group is confused. They must be clearly separated from microaerophilic and carbon dioxide requiring cocci which in the past have often been incorrectly considered anaerobic. One simple differential test is sensitivity to metronidazole: truly obligate anaerobic cocci are metronidazole-sensitive, the others are resistant.

Several genera have been described: *Veillonella* (minute Gram-negative cocci), *Peptococcus* (clustering Gram-positive cocci), *Peptostreptococcus* (chaining Gram-positive cocci), but their validity has been questioned because of variability in the results of Gram staining and the morphological appearances both microscopic and colonial.

Habitat: commensals in the oropharynx, colon and female genital tract.

Pathogenicity: local sepsis in mixed infections with other anaerobes (bacteroides and fusobacteria) and aerobes.

20

Lactobacillus

Members of this genus are widely distributed as saprophytes in vegetable and animal material (e.g. milk, cheese): others are common human and animal commensal parasites. Classification is complex and in dispute: characteristically species attack carbohydrates to form abundant acid and are tolerant of an acid environment (pH 3.0–4.0). The best recognised species is *Lactobacillus acidophilus*. Strictly anaerobic lactobacilli are now placed in a separate genus, bifidobacterium.

Lactobacillus acidophilus
Habitat: present in large numbers in the mouth, gastrointestinal tract and female genital tract; most vaginal lactobacilli (Döderlein's bacilli) seem to be *Lacto. acidophilus*.

LABORATORY CHARACTERISTICS

Morphology and staining: large, often thick (1–5 × 1 μm) Gram-positive bacilli with a tendency to form pairs in line, short chains and filaments; non-motile, non-sporing.

Culture: grow best under microaerophilic conditions in the presence of 5 per cent CO_2 and at pH 6.0. Small colonies form on ordinary agar after 48 h incubation: growth is better in media enriched with glucose or blood.

Selective media: acid media, e.g. tomato juice agar (pH 5.0) support the growth of lactobacilli but inhibit many other bacteria.

PATHOGENICITY

Lactobacilli are associated with dental caries.

21

Legionella

A recently discovered genus of unusual, even unique properties: the base composition of its DNA is distinct from that of all other bacteria. In addition, there are legionella-like organisms with similar biological characteristics but differing genetic composition.

Legionella pneumophila

Habitat: an environmental organism found in soil and water (including domestic water supplies and air conditioning units).

LABORATORY CHARACTERISTICS

Morphology and staining: slender rods or cocco-bacilli: Gram-negative but legionellae do not take up the counterstain particularly well.

Culture:

1. *On artificial media:* a fastidious bacterium difficult to grow on ordinary laboratory media: use either blood agar or charcoal yeast extract agar supplemented with L-cysteine and ferric pyrophosphate.
   Incubate for 21 days at 35–37°C in 5 per cent carbon dioxide; colonies usually appear in 3–5 days.
   Examine for growth and if suspect colonies are Gram-negative bacilli identify by: production of a fluorescent brown pigment on a special solid medium, biochemical reactions, direct or indirect immunofluorescence.
2. *In the fertile hen's egg:* this is the most sensitive culture medium.
   Inoculate: the yolk sac of a 6-day-old embryo.
   Observe: eggs that die 4–10 days later.
   Examine: the yolk-sac contents by direct or indirect immunofluorescence; culture on solid media for isolation.
3. *Animal inoculation*
   Inject: guinea pigs intraperitoneally.

Observe: signs of illness — fever, ruffled fur, watery eyes — over 3–5 days.

Examine: the peritoneal exudate, spleen and liver in the same way as the yolk sac contents described above.

Isolation from contaminated clinical material: concentration before culture can be achieved by treatment with hydrochloric acid, pH 2, to which legionellae are resistant. The use of antibiotic-containing fluid-enrichment and solid-selective media is also recommended.

Note. Diagnosis is usually serological: isolation from clinical material is rarely attempted.

Antigenic structure: six serogroups have been recognised.

PATHOGENICITY

The cause of *Legionnaire's disease* — a severe form of pneumonia — and less serious respiratory disease, e.g. Pontiac fever.

ANTIBIOTIC SENSITIVITY

Sensitive to erythromycin, rifampicin.

22

Spirochaetes

Spirochaetes are helical organisms which share many properties with Gram-negative bacteria.

Habitat: most are free-living and non-pathogenic but a few are causes of important human disease.

LABORATORY CHARACTERISTICS

Morphology and staining: spirochaetes have a unique helical structure: a central protoplasmic cylinder is bounded by the cytoplasmic membrane and a cell wall of similar structure to that of Gram-negative bacteria (Fig. 2.4, Chapter 2). Between a thin peptidoglycan layer and the outer membrane run the *axial filaments,* varying in number with the spirochaete, which are fixed at the extremities of the organism. They constrict and distort the bacterial cell body to give rise to the typical helical structure (Fig. 22.1).

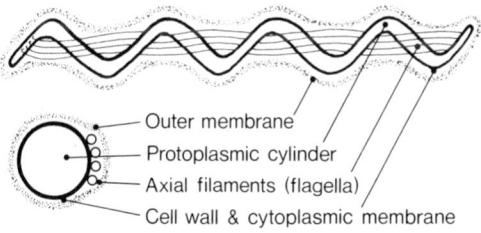

Outer membrane
Protoplasmic cylinder
Axial filaments (flagella)
Cell wall & cytoplasmic membrane

Fig. 22.1   The structure of a spirochaete.

The larger spirochaetes (e.g. *Borrelia species*) are Gram-negative: others stain poorly or not at all by the usual methods. They are too slender and too weakly refractile to be seen with the ordinary light microscope but can be rendered visible by dark-ground microscopy or by staining that enlarges their diameter, e.g. by deposition of heavy metals (Levaditi silver method) or by fluorescent antibody.

Motility is of three types:

1. Rotation about the long axis.
2. Flexion of the cells.
3. True movement, i.e. from one site to another.

Note. The axial filaments can be conveniently considered as inwardly turned flagella which do not extend into the external environment.

ANTIBIOTIC SENSITIVITY

Sensitive to penicillin and a number of other antibiotics; anaerobic spirochaetes are also sensitive to metronidazole.

Genera

Five are recognised:

1. Treponema ⎫   Species within these genera
2. Borrelia   ⎬   are associated with or pathogenic
3. Leptospira ⎭   for man and animals.
4. Spirochaeta — free living in the environment.
5. Cristispira — commensal, mostly in molluscs.

TREPONEMA

Treponema pallidum

Habitat: the disease lesions of primary and secondary syphilis.

Morphology: slender filamentous helices, 6–14 μm $\times$ 0.2 μm with 6–12 evenly spaced coils.

Culture: cannot be cultivated *in vitro* but can be propagated by inoculation of rabbits at certain sites, e.g. anterior chamber of eye, testes: this enables suspensions of *Tr. pallidum* to be prepared. These are used as the antigen to detect specific antibody in patient's serum.

Identification: in material from clinical lesions is by dark-ground microscopy.

Antigenic structure: all treponemes possess a common 'group' antigen; in addition, *Tr. pallidum*, *Tr. pertenue* and *Tr. carateum* share other antigens that make them similar but different from other treponemes.

PATHOGENICITY

Cause of the venereal disease syphilis.

Viability: a strict parasite that dies rapidly outside the body; it is very sensitive to drying.

Treponema pertenue

The cause of yaws, a chronic relapsing non-venereal treponematosis widespread in the tropics, which is characterised by ulcerative and granulomatous lesions in skin, mucous membranes and bone.

Treponema carateum

The cause of pinta, a non-venereal treponamatosis affecting dark-skinned people in Central and South America: the skin becomes hyperkeratotic and depigmented.

Tr. pertenue and *Tr. carateum* are morphologically indistinguishable from and antigenically very similar to *Tr. pallidum*. They cannot be cultivated *in vitro*.

Other treponemes

A number of species are found as commensals in the mouth, genital secretions and intestine. Their presence may create diagnostic problems when examining dark-ground preparations for *Tr. pallidum*.

Some can be grown *in vitro:* they are strict anaerobes.

BORRELIA

Borrelia vincentii

Habitat: the oro-pharynx, as a commensal and potential pathogen.

Morphology and staining: large spirochaetes 5–15 μm × 0.5 μm with irregular wide open coils; Gram-negative.

Culture: can be grown, with difficulty, in serum-enriched media; a strict anaerobe.

Identification: in exudates from clinical lesions by the appearance of a Gram-stained film.

Pathogenicity

In association with anaerobic fusiform bacilli responsible for gingivo-stomatitis and Vincent's angina.

Borrelia recurrentis

The cause of epidemic louse-borne relapsing fever.

Borrelia duttonii

The cause of endemic tick-borne relapsing fever.

Both diseases are encountered in parts of Asia, Africa and South America.

The diseases are characterised by febrile episodes alternating with afebrile periods and last for several weeks. Each relapse is the result of a change in the antigenic structure of the organism: antibodies already formed are ineffective against the new variants.

LEPTOSPIRA

Leptospira interrogans
Now regarded as the only species in the genus; it is divided into two complexes (interrogans and biflexa); 16 serogroups and more than 130 different serotypes are recognized.

Habitat. Many leptospira are free-living saprophytes (the biflexa complex); others cause disease in man and are parasites in the kidneys of some rodents and domestic animals (the interrogans complex).

Morphology: spiral organisms 5–20 μm $\times$ 0.1 μm with very numerous closely-set coils and hooked ends.

Culture: grow readily in serum enriched fluid or semi-solid media under aerobic conditions; optimum temperature is around 30°C.

Identification: the serotype of an isolated strain is determined by antigenic analysis.

Pathogenicity
Cause of the zoonotic disease leptospirosis.

Viability: pathogenic serotypes may survive for days outside the animal body in a moist environment as long as it is not acid.

23

Mycoplasma

Mycoplasmas are bacteria which lack cell walls. They resemble L-forms of bacteria but unlike them are independent naturally-occurring microorganisms.

Species: Table 23.1 lists some of the better-known mycoplasmas and their habitat.

LABORATORY CHARACTERISTICS

Morphology and staining: pleomorphic; several different forms exist varying from small spherical shapes — which can pass filters — to longer branching filaments; Gram-negative but stain poorly with Gram's stain; colonies on agar are best stained with Diene's stain.

Culture: on semi-solid enriched medium containing 20 per cent horse serum, yeast extract and DNA; incubate aerobically for 7–12 days with CO_2 or in nitrogen with added CO_2.

Observe for typical 'fried-egg' colonies embedded into the surface of the medium.

Note: T-strain mycoplasmas form minute colonies ('T' indicating tiny) and are now classified in the genus *Ureaplasma*; some mycoplasmas with less exacting growth requirements have been reassigned to a separate genus, *Acholeplasma*.

Identify by inhibition of growth round discs impregnated with specific antisera.

Table 23.1 *Mycoplasma species*

Species	Habitat
M. pneumoniae	Human respiratory tract
M. orale	Human mouth
M. salivarium	
M. hominis	Human genital, and possibly respiratory, tracts
Ureaplasma urealyticum	Human genito-urinary tract
Acholesplasma laidlawii	Soil, water

Antigenic structure: although there is some antigenic sharing in complement fixation tests, mycoplasma species are distinct in tests of growth inhibition. *Note:* mycoplasma are like viruses in that their replication is inhibited by specific antibody without complement; this is not the case with bacteria.

PATHOGENICITY

1. **Mycoplasma pneumoniae**
The only member of the group of which the pathogenicity is unequivocally established: it is a major respiratory pathogen responsible for two main syndromes:

(i) febrile bronchitis.
(ii) primary atypical or 'virus' pneumonia.

2. **Ureaplasma urealyticum (T-strain mycoplasma)**
Has been implicated in non-specific urethritis or genital infection but its role in this disease is not proven.

3. **Mycoplasma hominis**
The other genital species; may cause some cases of gynaecological sepsis.

ANTIBIOTIC SENSITIVITY

Sensitive to tetracycline — the drug of choice for treatment: also sensitive to aminoglycosides, erythromycin. Not surprisingly, mycoplasmas are resistant to antibiotics that interfere with bacterial cell wall synthesis, e.g. penicillin.

Tissue culture
Mycoplasmas are a common contaminant of cell lines, their persistence favoured by the presence of penicillin and some other antibiotics in the tissue culture media. The most frequently isolated species are *M. hominis* and *M. orale* — probably derived from the mouth of those handling the cells. *Acholeplasma laidlawii* is also occasionally isolated.

24

Candida

Candida are yeast-like fungi. Several species are found in man but one species, *Candida albicans*, is responsible for 90 per cent of infections. The other species include *C. stellatoidea* and *C. tropicalis*.

Candida albicans
Habitat: the normal flora of the upper respiratory, gastrointestinal and female genital tracts.

LABORATORY CHARACTERISTICS

Morphology and staining: two forms, both Gram-positive, are recognized in clinical material and on culture.

1. Spherical to oval budding cells (2–4 × 4–7 μm). The yeast or Y-form.
2. Elongated filamentous cells joined end-to-end and producing buds (blastospores). These pseudohyphae constitute the mycelial or M-form. *C. albicans* is the only species to produce abundant pseudohyphae *in vivo* (Fig. 24.1).

Culture: aerobic and easy to cultivate but isolation from clinical material may be impeded by faster growing bacteria.

1. Sabouraud's medium: a simple glucose peptone agar, pH 5.6, often made more selective by the addition of antibiotics (e.g. chloramphenicol) is useful for primary isolation. Incubation at 37°C for 48 h may be necessary
2. Ordinary agar and blood agar: colonies may be observed more easily around antibiotic discs which have inhibited bacterial growth.

Colonial morphology: Colonies are cream to white in colour, flat or hemispherical in shape and have a waxy surface. The yeasts are predominantly in the Y-form but M-forms develop in older cultures and the pseudohyphae may project from the edge of the colonies.

Identification: can be readily differentiated from other species by production of:

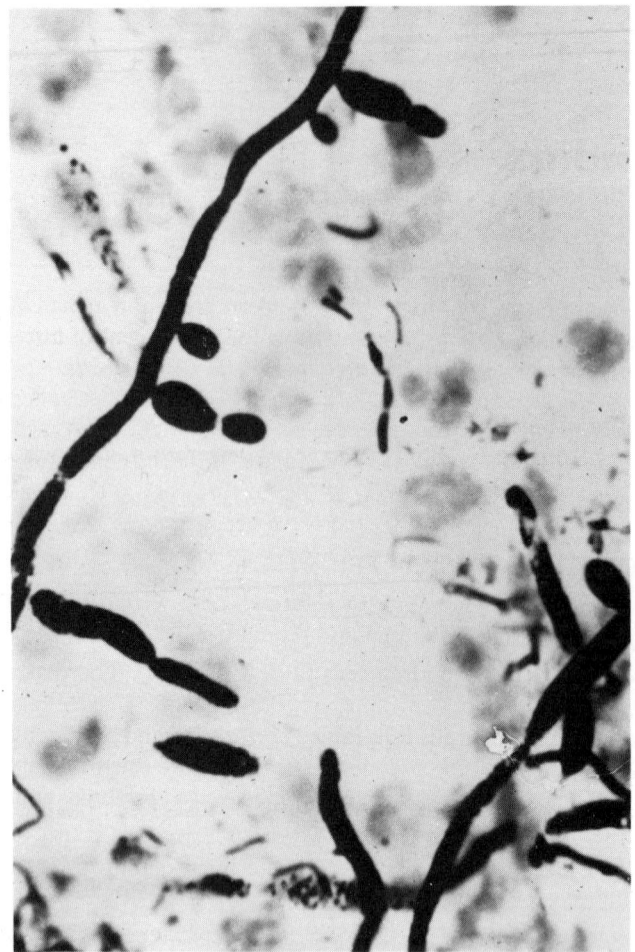

Fig. 24.1 *Candida albicans* in a film of exudate from a case of thrush showing pseudohyphae and budding cells (magnified × 4000).

1. *Germ-tubes*
   Method: grow in serum for 3 h at 37°C, make a wet film and examine for formation of filamentous outgrowths — germ tubes
2. *Chlamydospores*
   Method: grow in a nutritionally poor medium (e.g. cornmeal agar) for 24 h at 28°C and examine for presence of round thick-walled resting structures — chlamydospores — usually found at the ends of pseudohyphae deep in the agar

Biochemical activity: the results of fermentation (anaerobic metabolism) and assimilation (aerobic metabolism) of a range of carbohydrates are used in the identification of *Candida species.*

Antigenic structure: strains fall into two serotypes: A — antigenically similar to *C. tropicalis*; B — antigenically similar to *C. stellatoidea.* Candida antibodies can be demonstrated in most human sera. Delayed-type hypersensitivity is common and a positive Candida skin test is almost universal in normal adults.

PATHOGENICITY

Source: usually endogenous but cross-infection may occur, e.g. from mother to baby, from baby to baby in a nursery.

Host: infections are most common in babies who are premature and adults debilitated by general ill-health, notably diabetes. A special 'at-risk' group is composed of patients compromised either by the nature of their disease (e.g. malignancy, in particular leukaemias or lymphomas) or the treatment they have received (e.g. long courses of wide-spectrum antibiotics, immunosuppressive or cytotoxic drugs).

Lesions: infections (*candidosis*) are usually superficial (mucous membranes and skin) but occasionally are deep and involve internal organs.

Superficial
Mucous membranes: thrush — white adherent patches on buccal mucosa or vagina.

Skin: red weeping areas usually when skin is moist and traumatized, e.g. intertrigo in the obese.

Deep
Involvement of lower respiratory tract and urinary tract; septicaemia with localization in endocardium, meninges, kidney, bone.

ANTIBIOTIC SENSITIVITY

Candida are eukaryotic microorganisms and are resistant to *all* antibacterial antibiotics. They are sensitive to the polyenes (nystatin, amphotericin B), the imidazoles (miconazole, clotrimazole) and 5-fluorocytosine.

SECTION THREE

Bacterial disease

25

Normal flora

The normal human body has a profuse bacterial flora: this consists of *commensal* bacteria which, although parasitic, exist in a symbiotic equilibrium with the host. Many of the commensals are potential pathogens and if the balance is disturbed by some breach of the body defences or if a parasite of increased pathogenicity is acquired, infection may result. Nevertheless, virtually all 'unequivocal pathogens' may be encountered in healthy carriers.

Change and composition. The normal flora is not static: although a basic flora persists, it is subject to constant change and individual components wax and wane; the reasons for the abundance of certain bacteria at particular sites are not understood.

Beneficial role. The presence of the normal flora prevents other more pathogenic bacteria from gaining a foothold in the body. The gut bacteria seem to be responsible for the normal structure and function of the intestine: they degrade mucins, epithelial cells and carbohydrate fibre and their metabolism produces vitamins, especially vitamin K.

Alteration by antibiotics. Broad-spectrum drugs disrupt the composition of the normal flora by inhibiting sensitive organisms and allowing overgrowth of resistant bacteria. As a rule, the host can cope with these changes but they occasionally result in serious infection.

Bacteriologists require a detailed knowledge of the normal flora: most specimens cultured in the laboratory yield commensal bacteria and sometimes these are the pathogens responsible for the infection. Interpretation of the results of culture requires both knowledge and experience.

Distribution. With the exception of the alimentary tract, the internal organs of other systems are sterile in health, e.g. the bladder and kidneys, the bronchi and lungs, the CNS. Effective local defence mechanisms exist to maintain the sterility of these sites: in addition, chemical substances in serum and tissue fluids, e.g. complement, antibody, promote the powerful phagocytic activity of the polymorphonuclear leucocytes.

Below are the different sites of the body with the main bacterial species which make up their normal flora.

RESPIRATORY TRACT

The lower respiratory tract is sterile but the upper tract is colonised — heavily in the case of the mouth and nasopharynx (Table 25.1). Saliva contains about 10^8 bacteria per ml: gingival-margin debris and dental plaque consist almost entirely of microorganisms.

Table 25.1   The principal bacteria of the normal respiratory flora

Site	Bacteria	
Nose	Staphylococcus epidermidis Staphylococcus aureus Corynebacteria	
Oro-pharynx	Viridans streptococci Commensal neisseriae Corynebacteria Bacteroides	mainly B. melaninogenicus, B. oralis
	Fusobacteria ⎤ Spirochaetes ⎦ Lactobacilli Veillonella and other anaerobic cocci Actinomyces	Especially around the teeth
	Haemophilus influenzae ⎤ Streptococcus pneumoniae ⎦ Less common: Streptococcus pyogenes Neisseria meningitidis	The important potential pathogens

GASTROINTESTINAL TRACT

The oesophagus has a flora similar to that of the pharynx. The empty stomach is sterile due to gastric acid.

The normal flora of the duodenum, jejunum and upper ileum is scanty but the large intestine is very heavily colonised with bacteria (Table 25.2).

Faeces: Contain enormous numbers of bacteria which constitute up to one third of the faecal weight: the majority of these bacteria seem to be dead. The number of living bacteria is about 10^{10} per gram and almost all (99.9 per cent) are anaerobes: the anaerobic environment of the colon is maintained by the aerobic bacteria utilising any free oxygen. *Bifidobacteria* are Gram-positive bacilli similar to lactobacilli but strict anaerobes: like lactobacilli they are virtually non-pathogenic: they and *Bacteroides species* are the dominant anaerobes.

Bacteroides fragilis is a considerably rarer gut inhabitant than other species classified in the fragilis group but has much greater potential for pathogenicity.

Table 25.2   Bacteria of the large intestine

Bacteroides (mainly members of the fragilis group which outnumber *B. fragilis* itself)
Bifidobacteria
Anaerobic cocci
Escherichia coli
Streptococcus faecalis
Clostridia
Lactobacilli
Less common inhabitants:
   Klebsiella species
   Proteus species
   Enterobacter species
   Pseudomonas aeruginosa

GENITAL TRACT

For anatomical reasons the female genital tract is much more heavily colonised than that of the male. Normal vaginal secretions contain up to 10^8 bacteria per ml. The genital flora is shown in Table 25.3.

Table 25.3   Main bacteria of the male and female genital tracts

Female	
Vulva	*Staphylcococcus epidermidis*
	Corynebacteria
	Escherichia coli and other coliforms
	Streptococcus faecalis
	Yeasts
Vagina	Lactobacilli (known as Doderlein's bacilli)
	Bacteroides (especially *B. melaninogenicus*)
	Streptococcus faecalis
	Corynebacteria
	Yeasts
Male and female distal urethra	*Staphylococcus epidermidis*
	Corynebacteria

Note. The secretions of both male and female genitalia may contain *Mycobacterium smegmatis* — acid-fast bacilli which, if they contaminate urine specimens, can easily be mistaken for tubercle bacilli.

Mycoplasma. Strains of mycoplasma called T-strains or ureaplasma which form minute colonies on culture are commonly present as part of the normal genital flora of both sexes.

SKIN

Although the hard dry surface of the skin may seem at first sight to be less hospitable for bacteria than moist mucous membranes, the skin has a rich resident bacterial flora (estimated at 10^4 organisms per cm^2). It is not evenly distributed: the bacteria exist in microcolonies of 10^2–10^3 organisms.

Anaerobic organisms predominate — particularly in areas with many sebaceous glands where anaerobic conditions prevail. In moist skin e.g. the axilla and groin, coliform organisms are often present (Table 25.4).

Table 25.4   Main bacteria of the skin flora

Propionibacterium acnes	
Anaerobic cocci	
Staphylcococcus epidermidis	
Micrococci	
Corynebacteria	
Less common:	
Staphylococcus aureus	This potential pathogen is present in about 50 per cent of normal adults
Coliforms	

External auditory meatus

An extension of the skin and often profusely colonised: the main species found are *Staphylococcus epidermidis* and corynebacteria; acid-fast mycobacteria are occasionally present in the wax.

Conjunctival sac

Bacteria are scanty: occasional *Corynebacterium xerosis* and *Staph. epidermidis*.

26

Host-parasite relationship

Infection is the result of breakdown in the host-parasite relationship and follows when the balance is tipped in favour of the parasite.

Man - the host - lives in general balance with his environment. This environment includes the numerous bacteria found in all sites, animate and inanimate, with which man comes in contact: most important is his own normal flora.

Most bacterial disease is due to organisms which constantly — or at least from time to time — form part of the commensal flora: naturally there are exceptions and some pathogenic bacteria are found only in disease.

Selection pressure favours the survival of bacteria with limited pathogenicity which maintain a symbiotic relationship with their host. Virulent pathogens which severely incapacitate or kill man are denied the opportunity to spread within a community because their host is no longer able to circulate and come in contact with other individuals: human beings are gregarious but only when they are healthy.

DEFENCE MECHANISMS OF THE HOST

A number of host factors influence the outcome of host-parasite interaction. Some are linked with socioeconomic status, e.g. malnutrition is associated with poverty and overcrowding, conditions that also favour the transmission of a virulent pathogen within a community. Some of these host factors are listed below.

1. *Nutrition:* malnutrition predisposes to infection.
2. *Age:* the very young and the aged are particularly liable to infection.
3. *Sex:* differences in the male : female incidence of infection are small and sometimes can be explained by differences in the liability to encounter the pathogen, e.g. occupational risks.
4. *Race:* e.g. negroes are more susceptible than whites to tuberculosis.
5. *Occupation:* some occupations (e.g. those associated with inhalation of minerals) have a higher than normal risk of infection with certain microorganisms.

6. *Impairment of the host response by:*

   a. Treatment e.g. immunosuppressive, cytotoxic or steroid drugs, radiotherapy

   b. Disease e.g. malignancy (especially leukaemia or lymphoma): metabolic diseases (diabetes, renal or hepatic failure)

The host has a number of defence mechanisms with which to counteract bacterial aggression. These can be divided into two categories:

1. *Non-specific:* in which the defence mechanism is not directed at a particular organism
2. *Specific:* in which the defences of the host are directed against a particular organism; these are dependent on immunological mechanisms.

NON-SPECIFIC DEFENCE MECHANISMS

1. *Skin*

Skin is a tough layer or integument which forms an impermeable barrier to invasion of the tissues below by organisms from either the normal flora of the skin or the environment: the frequency of infection when this barrier is breached, e.g. by a surgical or traumatic wound, illustrates the efficiency of the skin in preventing infection.

2. *Normal flora*

Present at many sites, makes it difficult for pathogens acquired from without to establish themselves. Fatty acids with antibacterial activity are produced by skin flora from glycerides in sebum and by intestinal anaerobes from the contents of the colon.

3. *Lysozyme*

In tears and other body fluids lysozyme is an enzyme which lyses the mucopeptide of the cell wall of Gram-positive bacteria.

4. *Flushing action of:*

   Tears: which with lysozyme keep the surface of the eye sterile.

   Respiratory tract mucus: which traps bacteria and constantly moves upwards propelled by the *cilia* on the cells of the epithelium.

   Urine: voiding clears bacteria that may have gained entry to the bladder.

5. *Low pH*

   Stomach acid: ingested bacteria are usually destroyed by the low pH of the stomach contents; this can, of course, be buffered by the presence of food.

Vaginal secretions have an acid pH during the reproductive years due to lactobacilli which metabolise glycogen present in the epithelium because of circulating oestrogens; the lactic acid produced helps to prevent access of harmful bacteria.

PHAGOCYTOSIS

Phagocytosis is the most powerful — and therefore the most important — of the defence mechanisms: it is mediated by scavenger cells which ingest invading organisms and destroy them intracellularly by the action of enzymes.

There are two types of phagocyte:

1. Neutrophil polymorphonuclear leucocytes (the 'polymorphs')
Also known as *microphages*. They are produced in the bone marrow and, when mature, circulate in the bloodstream for 6 –7 h. They are short-lived cells which arrive rapidly at the scene of infection attracted by chemotactic substances elaborated during the inflammatory process. Phagocytosis is promoted by specific antibody and complement which act as *opsonins*. Polymorphs act as an early defence against infection and are the 'pus cells' seen in the exudate from acute infections.

2. Macrophages of the mononuclear phagocytic (or *reticulo-endothelial*) *system*
Produced in bone marrow, they travel as monocytes in the blood stream to become distributed as *free macrophages* (in lung alveoli, the peritoneum and inflammatory granulomas) or *fixed macrophages* integrated into the tissues (in lymph nodes, spleen, liver (Kupffer cells), CNS (microglia) and connective tissue (histiocytes)). Phagocytosis by these long-lived cells can either be non-specific or promoted by antibody and complement. Macrophages process bacterial antigens and present them to lymphocytes in such a way as to stimulate a specific immune response: they also play an important part in cell-mediated immunity.

Phagocytic function can be divided into four stages:

1. *Chemotaxis:* attraction of the phagocyte to the site of the organism.
2. *Attachment* (*adherence*) of the bacterium to the membrane of the phagocyte.
3. *Ingestion* in which the phagocytic cell extends small pseudopods to envelop the bacterium: these fuse to form a pouch or *phagosome*. Lysosomes containing hydrolytic enzymes and other bactericidal substances migrate towards the phagosome and fuse with its membrane to form a *phagolysosome*.

4. *Intracellular killing of the ingested bacterium:* most bacteria are killed within a few minutes of phagocytosis although the degradation of the bacterial cell may take several hours.

COMPLEMENT

A family of at least nine proteins present in serum: these react together one after another in a cascade once the reaction has been triggered. Activation of the first stage takes place when specific antibody combines with a bacterial or other antigen: the sequential reaction liberates fragments that attract phagocytic cells (chemotaxis), promotes subsequent phagocytosis and induces the changes characteristic of the inflammatory reaction.

THE IMMUNE SYSTEM IN INFECTION

Below is an outline of the immune response against bacterial infection.
    There are two main mechanisms:
1. *The humoral (antibody) response*
2. *The cell-mediated response*

1. ANTIBODY RESPONSE

Antibodies are proteins in the blood stream which are produced in response to infection by a microorganism: they are specifically directed against the bacterium — or its component parts. Bacterial components are usually protein — occasionally carbohydrate — and since most are *antigens* they stimulate antibody production.
    *B-(bone marrow derived)*lymphocytes. When an antigen — in this context a bacterium — encounters B-lymphocytes in the spleen or lymph nodes the lymphocytes are activated and changed into antibody-secreting plasma cells (the classical cells of humoral immunity). The antigen may be presented by macrophages and involvement of T-lymphocytes is required to initiate the immune response to some antigens.
    Antibodies are *immunoglobulin* (Ig) protein molecules: their structure is Y-shaped and consists of a *Fc fragment* (the stem of the Y) and two *Fab fragments* (the arms of the Y): the Fab fragments contain the combining sites for specific antigens and, in antibodies to different antigens, show highly variable amino acid sequences. The Fc fragment of different antibodies, on the other hand, has a relatively constant amino acid composition and is the site for attachment of complement.

Although there are five types of immunoglobulin only three are concerned in the response to infection. These are described below:

1. *IgM* — a pentamer of IgG of 1×10^6 molecular weight — the first antibody produced: appears approximately a week after infection and persists for about 4 to 6 weeks.
2. *IgG* — 1.6×10^5 molecular weight monomer — the main antibody produced: appears about 2 weeks after infection but persists for long periods of time — *late antibody*.
3. *IgA* — 1.7×10^5 molecular weight, a monomer in blood, present as a dimer in body secretions e.g. saliva, respiratory and alimentary mucus, tears, colostrum, etc. IgA in extracellular fluids — secretory IgA — is coupled to a carbohydrate transport piece which is not found on serum IgA.

Action

Antibodies are powerful defence mechanisms against viruses since they neutralise viral infectivity: they are much less effective on their own (i.e. without complement) against bacteria.

Nevertheless antibodies are important in combating bacterial infection principally by the following specific actions:

Neutralisation of toxins

Promotion of phagocytosis: antibody-coated bacteria are more readily phagocytosed, i.e. they are opsonised

Bacterial lysis: certain Gram-negative bacilli are lysed in the presence of antibody and complement.

2. CELL-MEDIATED IMMUNITY

Delayed hypersensitivity — or cell-mediated immunity — was first described in tuberculosis in the late nineteenth century: the mechanism, however, has only recently been discovered. Delayed hypersensitivity is particularly important in infections due to organisms which persist or multiply intracellularly such as those causing tuberculosis, leprosy, brucellosis and virus infections. In delayed hypersensitivity there is infiltration of the inflammatory lesion with sensitised T-lymphocytes.

T-(thymus dependent)lymphocytes are a population of lymphocytes developed in the thymus. Responsible for cell-mediated immunity they comprise 75 per cent of the circulating lymphocytes in man: when sensitised or primed T-lymphocytes encounter the specific antigen with which they can react they undergo transformation to become actively metabolising blast cells with release of lymphokines.

Lymphokines have the following activities:
Inhibition of macrophage migration (this probably localises the macrophages to the site of infection)
Chemotactic attraction of lymphocytes, macrophages and polymorphonuclear leucocytes to the site of infection
Increase in capillary permeability
Mitogenic activity: or stimulation of normal, i.e. unsensitised lymphocytes to divide
The overall effect of delayed type hypersensitivity is to limit the size of the lesion and the surrounding inflammatory reaction and to localise the organism within it.

AGGRESSIVE MECHANISMS OF THE PARASITE

Bacteria vary in their *pathogenicity* — or ability to produce disease — in man.
Virulence is a commonly used but ill-understood term which indicates the degree of pathogenicity.
Neither pathogenicity nor virulence are easy to measure: it is impossible to do so in human beings for ethical reasons; experiments in laboratory animals which measure the incidence of disease or death following inoculation of an organism are not always — and probably not often — analogous to the behaviour of the organism in the human host.
Bacteria as pathogens have two basic mechanisms of producing disease:

1. *Invasiveness*
2. *Production of toxins*

Although bacteria can cause disease which is predominantly invasive or toxic in origin, most bacterial infections are due to a combination of both activities.

Invasiveness
Invasiveness is the ability of an organism to spread within the body once it has gained its initial foothold. This depends on (i) the action of toxins elaborated by the bacterium and (ii) cell surface components which enable it to resist phagocytosis. The latter may be demonstrable as visible capsules (e.g. the pneumococcus) or present as part of the cell wall (e.g. the M protein of *Strep. pyogenes*, the K antigens of the enterobacteria).

A high degree of bacterial invasiveness is associated with severe infection: spread from a local site is often via the lymph channels (*lymphangitis*) to the draining lymph nodes (*lymphadenitis*) and possibly from these to the blood stream with consequent *septicaemia*: this is one of the most serious manifestations of infection.

Bacterial toxins

These are of two types:

1. *Exotoxins*

Liberated extracellularly from the intact bacterial cell (although also produced on cell lysis) exotoxins spread via the blood stream or sometimes, nerves: can produce ill effects locally and also at sites far distant from the infective process. A few bacterial diseases (e.g. diphtheria, tetanus) are the result of microorganisms localised at the site of entry forming exotoxins which produce severe, distant characteristic effects.

2. *Endotoxins*

Endotoxins are O antigens, structural components of the cell wall of Gram-negative bacteria; liberated only on cell lysis or death of the bacterium. Although differing in antigenic structure, they all produce the same physiological effects.

Some differences between exotoxins and endotoxins are shown in Table 26.1.

Table 26.1   Bacterial toxins

	Exotoxins	Endotoxins
Composition	Protein	Lipopolysaccharide
Action	Specific	Non-specific
Effect of heat	Labile	Stable
Antigenicity	Strong	Weak
Produced by	Gram-positive bacteria (few exceptions)	Gram-negative bacteria (no exceptions)
Conversion to toxoid*	Yes	No

toxoid is toxin treated, usually with formaldehyde, so that it loses toxicity but retains antigenicity.

Some bacteria produce or contain various substances often called 'aggressins' (in addition to toxins directly harmful to the host) which enable them to withstand the host defences:

phospholipase — cytolytic
coagulase   — deposits fibrin
hyaluronidase — dissolves cell-binding material, aids spread

CONCLUSIONS

The host–parasite relationship is complex and delicately balanced: the defence mechanisms of the host protect against invasion or uncontrolled replication of bacteria but allow a symbiosis with the normal flora.

The parasite, for its part, depends for long-term survival in human populations on not harming the host or at least causing only minimal damage.

When this balance is disturbed internally, or a result of invasion from outside, bacterial disease results.

27

Epidemiology

Epidemiology is the study of the spread of infection. From the point of view of the origin of the infection, bacterial disease can be considered in two categories:

1. *Endogenous:* when the organism is derived from the individuals own flora; but *note* – an epidemic organism may first be acquired as part of the normal flora.

2. *Exogenous:* when the organism is acquired from outside sources.

Epidemiology is chiefly concerned with exogenously-acquired infection and how infectious disease affects a community or the population at large.

RESERVOIRS AND SOURCES

Often, probably in most instances, the reservoir and the source of infection are one and the same: but not always, because sometimes the source has acquired the infecting organisms from the reservoir — the reservoir remaining distant from the eventual victim of the infection.

Reservoirs and sources are numerous. In terms of epidemiology the principal reservoirs and sources of communicable or infectious disease are as follows.

1. Human beings
By far the most important source of infection. There are three main ways in which human beings act as sources of infection to others.

i. *Active cases of disease*
Patients suffering from an infectious disease often shed the causal organism in large numbers; on the other hand, they may be so incapacitated by illness that they are prevented from circulating in the community.

ii. *Inapparent (subclinical) infections*

People suffering from the disease but without symptoms are important sources of infection; they continue to circulate in the community and, unless laboratory tests are carried out, are unrecognised as sources of infection.

iii. *Carriers*

Some patients — some of whom may have had symptomless infections — become long-term, even permanent, carriers and excretors of epidemic organisms; in some instances these organisms have become part of the normal flora, in others they are shed from some chronic, often symptomless, focus of infection; carriers, of course, also circulate in the general population and are usually undiagnosed.

2. Animals

Infection is a particular occupational hazard to farm workers, veterinary surgeons and slaughtermen.

Animal products such as meat, milk, hides, etc. can also be sources of infection to the general population.

3. Food

A most important source of infection; it may be infected either at its animal origin (e.g. poultry, milk) or later when handled by man. It acts as more than a passive vehicle; pathogenic bacteria multiply rapidly and in some instances produce toxins if the contaminated food is allowed to remain at the environmental temperature.

4. Water

Britain has a safe and pure water supply which is uncontaminated by sewage. If this sewerage system breaks down or in countries (which include most of those in tropical or sub-tropical areas) where the water supply is often polluted, large-scale water-borne outbreaks of infection can ensue; naturally, these outbreaks are usually due to faecal bacteria.

5. Soil

Most organisms in the soil are free-living bacteria which are either non-pathogenic or of very low pathogenicity; however, soil may become contaminated with pathogenic organisms derived from animal faeces or discharges.

6. Air

Air has a resident flora which mostly consists of bacteria of relatively low pathogenicity; however, air bacteria are also derived from human beings, who shed organisms both from skin as desquamated scales and from the respiratory tract; air also contains dust particles.

7. Dust
Dust is also contaminated with bacteria shed from human beings and from clothing, furnishings, bed linen and so on; bacteria in the air (particularly if contained in droplets) are deposited or fall into the dust and, conversely, dust particles are swept up into the atmosphere by air currents or movement of personnel.

8. Fomites
These are, strictly speaking, objects of a porous nature which absorb and can pass on contagion: in practice, any object which can be contaminated with bacteria is regarded as a fomite.

It can be seen that many of these reservoirs and sources are interrelated and interdependent.

ROUTES OF INFECTION
The route of infection largely depends on the reservoir or source. The main routes are listed below.

1. Inhalation
Inhalation especially of droplets of infected respiratory secretions from someone suffering from active infection or a carrier.

Droplets produced by sneezing and coughing are either:

(i) *Large* — in which case they travel only a few feet and contaminate the environment to produce infected dust and fomites.

or

(ii) *Small* — so that they evaporate to droplet nuclei (5 μm in diameter or less) and remain suspended in the air; these droplets may be inhaled to reach the lower respiratory tract directly.

2. Ingestion
A very common mode of spread: organisms derived from faeces may be passed on by the classic *faecal–oral route* usually via contaminated fingers, towels, etc.: faecal organisms may also be spread indirectly via contaminated drinking water or food.

3. Contact
Infection said to be spread 'by contact' is in fact most often due to inhalation of respiratory sections. Direct contact is of a close personal nature, e.g. via hands, kissing; indirect contact with spread from fomites usually involves final tranfer via the hands.

4. Sexual transmission
Several infectious diseases are spread wholly or partially as a result of sexual intercourse; not surprisingly, such diseases usually consist of lesions on the genitalia, and the organisms — which often do not survive well outside the body — are inoculated directly onto genital mucous membrane.

5. Inoculation
Directly through broken skin either due to accidental trauma or as a result of surgery is a common route of infection; infection can also be introduced by medical procedures such as catheterization, injection, blood or other transfusions, etc.

6. Vector-borne
Arthropod insects and parasites such as mosquitoes, ticks, lice and fleas — especially if they bite and are blood-sucking — are a route of infection; common in tropical countries but rare in temperate countries like Britain.

7. Transplacental
A few maternal infections can cross the placenta to infect the fetus, notably syphilis (a bacterial disease) and viral diseases such as rubella or cytomegalovirus infection.

EPIDEMIOLOGY

In order to control or prevent infectious disease, doctors must understand its reservoir, source and route of transmission: sometimes even when these are known it is impossible to control, for example, surgical wound infection, food-poisoning, most respiratory infections.

The spread of an infectious disease in a community depends on several factors.

1. *Number of susceptible hosts*
The chance of an outbreak of infection correlates directly with the number of susceptible people: immediately after an outbreak the general level and incidence of antibody is high, i.e. there is good *herd immunity*; with time, the level of antibody wanes and more children — who have never experienced the infection (and therefore have no antibody at all) — are born into the population. The herd immunity then becomes low with a relatively large number of susceptible hosts;

under these circumstances the organism can infect again on a large scale.

Geometric mean titre of antibody to a particular organism within the population is an indication of herd immunity: this is calculated from the mean of the logarithms of the titres in the serum samples tested; this method of calculation removes the bias created by one exceptionally high (or exceptionally low) titre of antibody in the sample.

2. *Pathogenicity*
Some organisms inherently possess a high capacity to spread (infectiousness or communicability) or to cause disease (virulence). Some organisms — like *Salmonella typhi* — can infect in very small doses: others — like *S. typhimurium* — require large numbers of organisms to establish infection; *S. typhi* is therefore more virulent than *S. typhimurium*. Influenza virus is the classic example of a microorganism with an extremely high infectiousness.

3. *Route of spread*
This factor may determine an outbreak of infection: for example, if there is a breakdown in the pure water supply, faecal organisms from sewage may be given an opportunity to infect large numbers of people. The increase in sexual promiscuity is responsible for the epidemic of gonorrhoea in recent years.

4. *Carriers*
When there is a relatively high proportion of carriers in a community this also increases the likelihood of an outbreak of infection. Outbreaks of meningococcal meningitis are often preceded by an increase in the proportion of people in the community who carry the organism in their throat.

5. *Climate*
Climate may contribute to the incidence of infection. Cold wet weather increases the number of exacerbations of chronic bronchitis whereas hot weather enhances the risk of food-poisoning.

MEASUREMENTS IN EPIDEMIOLOGY

Epidemiologists have various measurements by which infection in a community can be assessed: a community may be a family, an institution, a geographical area or the population of an entire country.

These measurements include:

1. *Incubation period*

To trace the spread of an outbreak it can be of great importance to know the incubation period — particularly if this is relatively long (e.g. about 2 weeks or longer): this may also allow time for preventive or containment measures to be instituted.

2. *Incidence or incidence rate*

This is the number of cases of disease in the community expressed as the ratio of the number of cases per 1000 (or 10 000, 100 000 or per million — whatever is appropriate) people in the population concerned; *prevalence* is a similar estimate but usually refers to the incidence in a population within a certain stated time.

3. *Attack rate*

Another way of expressing incidence: the term is used when the rate is applied to a particular defined group, for example, inhabitants of an institution.

4. *Secondary attack rate*

An important statistic for an epidemiologist, this is the number of secondary cases of infection which appear in the contacts – e.g. family or workmates, etc. – of an index case of the disease.

5. *Mortality rate*

The proportion of people in the community who die from the disease: again, this is expressed as deaths per 1000, 100 000, etc.

6. *Case fatality rate*

This rate, on the other hand, is the proportion of patients with the disease who die as a result of it: this is usually expressed as a percentage.

SURVEILLANCE

Measurements of infectious disease are constantly being monitored in communities with a good public health organisation. Although outbreaks cannot always — perhaps seldom — be prevented, in some cases appropriate measures can successfully halt the spread of an epidemic disease.

EPIDEMIOLOGY OF BACTERIAL DISEASE

Most worldwide *epidemics* or *pandemics* nowadays are viral but cholera is an exception and the El Tor variety of cholera is now spreading west from its original focus in the Far East: countries with adequate systems for sewage disposal are not at risk from this disease.

Many bacterial diseases have a constant *high incidence* in the population and are a major problem for public health authorities. These include salmonella food-poisoning, gonorrhoea and dysentery — although the latter disease is at present waning after a prolonged period of high incidence which has lasted for over 20 years. The reservoirs, sources and routes of spread of these diseases are well understood but it is impossible to control them.

Other bacterial diseases like streptococcal sore throat (which, for example, increased sharply in incidence in Glasgow during the winter of 1979-80) are always endemic in the population and wax and wane — usually for no apparent reason.

Immunisation can radically alter the epidemiology of an infectious disease — indeed, the main purpose of immunisation is to do this. Vaccines are generally less effective in the prophylaxis of bacterial diseases than of viral diseases. *Toxoids* are an exception to this and diphtheria and tetanus prophylaxis by toxoids is extremely effective.

Whooping cough vaccination is a controversial subject at present which illustrates a problem common to other bacterial diseases also. The vaccine has a protective effect but this is incomplete. Unfortunately it may also have serious side effects and can occasionally cause convulsions and brain damage. Although the incidence of these side effects is low, widespread publicity about some tragic cases resulted in a decline in the acceptance rate of triple vaccine: the widespread outbreak of whooping cough in 1978-79 was almost certainly at least partly a result of this.

Rare outbreaks such as that of typhoid fever in Aberdeen in 1964 — which was due to a contaminated tin of corned beef — illustrate the need for contant vigilance by public health authorities: reports in early 1980 of an outbreak of tuberculosis amongst school children in Manchester are a reminder that this disease too is far from being eradicated in Britain today.

28

Specimens for bacteriological investigation

Laboratory diagnosis in bacteriology depends on:

1. Isolation
2. Direct demonstration of the causal organism
3. Serology

1. ISOLATION

This depends on culture of material from the site of infection: obtained by:

(i) Collection of exudate, discharge, excreta or infected fluid into screw-capped sterile containers.

(ii) Sampling the infected area by a swab, i.e. a wooden or metal shaft with a cotton-wool tip rubbed on or inserted into the lesions and replaced in a stoppered plastic or glass tube for transport to the laboratory.

Specimens
Some examples are given below

1. *Urine:* a mid-stream specimen
   Delay: if this is inevitable, store the sample at 4°C or use a container with boric acid to prevent bacterial mutiplication.
   Dip slide: a slide coated with medium is dipped into the urine and replaced in a sterile container for transport to the laboratory; delay is unimportant since the organisms are already inoculated onto solid medium.
   Tuberculosis: if suspected, collect three entire first morning specimens.
2. *Faeces:* collect in a plastic container; if not available, a rectal swab can be taken.
3. *Sputum:* a morning specimen in a wide-mouthed container.
   Tuberculosis: if suspected, specimens collected on three consecutive mornings.

4. *Serous fluids:* (e.g. pleural, synovial, ascitic fluids); collect in a sterile container with citrate to prevent clotting.
5. *Cerebrospinal fluid:* collect by lumbar puncture into a sterile container.
6. *Blood culture:* blood, aseptically collected, is injected into each of two screw-capped bottles with a perforation in the cap to allow injection through the rubber liner; one bottle contains aerobic, the other anaerobic, medium.

Swabs
Widely used and essential for sampling some areas, e.g. throat, cervix; where *pus* is available it is always better to collect it for examination.

Transport
Most specimens need to be sent to the laboratory without delay: some bacteria die off quickly outside the body, others may overgrow and give a false impression of their original numbers.

If delay cannot be avoided specimens should be kept cool or at room temperature, except blood cultures and CSF which should be incubated at 37°C.

Stuart's transport medium: a sloppy agar containing salts with a reducing agent and sometimes pieces of charcoal: used with swabs: when swabs are placed in tubes containing this medium, delicate organisms are preserved.

2. DIRECT DEMONSTRATION
Some diseases can be diagnosed simply by demonstrating the causal organism morphologically in a stained smear or sometimes in an unstained wet preparation. This is rarely possible with material collected on a swab — a specimen of pus, exudate or other fluid or tissue should be sent to the laboratory to do this.

Immunofluorescence. Demonstration of the causal organism by direct or indirect immunofluorescence not only detects its presence in the material under examination but also identifies it serologically.

3. SEROLOGY
Specimen: 5–10 ml clotted blood in a sterile container.

Labelling
All specimens must be labelled with the patient's name and ward (or home address). Other details, including the nature of the specimen, clinical history, antibiotic therapy, etc., are essential for the interpretation of results.

Safety

Specimens which may present a hazard to laboratory staff (e.g. blood samples positive for hepatitis B virus, sputum from a known case of open pulmonary tuberculosis, etc.) must be labelled 'DANGEROUS SPECIMEN'.

Doctors must remember that they too may infect themselves when taking a specimen unless this is done carefully.

29

Respiratory tract infections

A very important cause of sickness — reckoned to account for a half of general-practitioner consultations and a quarter of all absences from work due to illness.

Route of infection — inhalation. Control of infections acquired by inhalation of infected secretions is well-nigh impossible. More frequent in winter time (October to March), close contact in school, at work and socially allows ready transfer of the causal agents — 'Coughs and sneezes spread diseases'. In family outbreaks infection is often introduced by the most susceptible member, usually a pre-school or school-age child. Immunisation is only available against a few specific infections, e.g. diphtheria, whooping-cough, influenza.

Respiratory infections can be classifed into four groups:

1. Infections of throat and pharynx
2. Infections of middle ear and sinuses
3. Infections of trachea and bronchi
4. Infections of the lungs

1. INFECTIONS OF THROAT AND PHARYNX

CLINICAL FEATURES

Sore throat is the commonest symptom accompanied by a variable degree of constitutional upset. Typical throat appearances for the different aetiological agents are described but it is often impossible to decide on the cause of a sore throat by clinical examination alone. Over two-thirds of these infections are caused by viruses — often with sore throat as part of the common-cold syndrome: the remainder are bacterial in origin almost all due to *Streptococcus pyogenes.*

Streptococcal sore throat

Mild redness of tonsils and pharynx may be the only sign but the classical picture is of injection and oedema involving the fauces and soft palate with exudate — *acute follicular tonsillitis.*

In severe cases this may be complicated by a peritonsillar abscess (*quinsy throat*) and extension of the infection to involve the sinuses and middle ear producing *sinusitis* and *otitis media*. Systemic illness with fever is the rule and the cervical lymph nodes may be enlarged. *Scarlet fever* is a streptococcal infection — usually a sore throat — accompanied by an erythematous rash when the infecting strain of *Strep. pyogenes* produces erythrogenic toxin in a susceptible (i.e. non-immune) patient — usually a child.

Incubation period: 1–3 days.

Source: cases or carriers. After an acute attack transient carriage for a few weeks is common. Throat carriers outnumber nasal carriers but the latter, who often have an associated sinusitis, are much more effective disseminators.

Treatment: penicillin is the drug of choice. Therapy should be started parenterally and continued orally for 10 days to prevent complications and further spread of the organism to contacts. Patients hypersensitive to penicillin are given erythromycin: tetracycline resistance is common (about 20 per cent of strains) and tetracycline is therefore contraindicated.

Late complications of streptococcal infections
Streptococcal infections, usually sore throat, are sometimes followed by disease which appears to be immunologically induced. The disease is of two main kinds:

1. Rheumatic fever
2. Acute glomerulonephritis

Rheumatic fever
Clinically: the acute onset of fever, pain and swelling of the joints and pancarditis, on average 2 to 3 weeks after streptococcal sore throat. The most serious manifestation is involvement of the heart: patients commonly have myocarditis, sometimes in addition pericarditis and endocarditis. The disease has been said 'to lick the joints but bite the heart'.

Now relatively uncommon it was formerly a disease associated with poor living conditions and overcrowding — circumstances that facilitated the spread of streptococcal sore throat.

Prognosis. Rheumatic fever usually clears up spontaneously although it has a marked tendency to recur: the main problem is that after the acute phase of the disease, patients later — often much later — develop chronic valvular disease of the heart, usually stenosis or incompetence of the mitral or aortic valves.

Pathology. Rheumatic fever is a disease of connective tissue which is almost certainly immunological in origin. The typical lesion is the *Aschoff nodule* — a pale hyaline focus with lymphocytic and macrophage infiltration and sometimes giant cells.

Immunology. Rheumatic fever may be the result of antibodies produced against protein and polysaccharide cell wall antigens of *Strep. pyogenes* cross-reacting with myocardial and heart valve tissue.

Laboratory diagnosis
The diagnosis can normally be made clinically but it is useful to check for the continuing presence of *Strep. pyogenes* in the throat or serological evidence of recent infection.

Isolation
Specimen: throat swab.
Culture: blood agar aerobically and anaerobically.
Observe: for β-haemolytic small dry colonies.
Identify: by Lancefield grouping.

Note. Rheumatic fever may follow infection with any serotype (Griffith type) of *Strep. pyogenes.*

Serology
Specimen: blood.
Examine: for antibody to streptolysin O (ASO) — a haemolysin produced by *Strep. pyogenes*: an ASO titre of 200 units or more is regarded as significant and sera from about 75 per cent of patients with rheumatic fever give a positive result. Tests are available to detect antibody to other streptococcal products, e.g. hyaluronidase, DNAase: they may be positive when the ASO titre is not raised. By using one or more of these serological tests it is normally possible to indicate that there has been an antecedent streptococcal infection.

Use of antibiotics
Antibiotics are required only to eradicate *Strep. pyogenes* from the throat: prophylactically on a long-term basis they are mandatory to prevent reinfection with risk of precipitating a recurrence: penicillin is the drug of choice.

Acute glomerulonephritis
Also a complication — probably of immunological origin — which may follow streptococcal throat infection or, occasionally, impetigo. Unlike rheumatic fever, it is particularly liable to follow infection with certain serotypes of *Strep. pyogenes* notably, in throat infections, type 12 (the main nephritogenic streptococcus) and, in skin infections, type 49.

Clinically: acute glomerulonephritis presents 1-5 weeks after a streptococcal throat infection with haematuria, albuminuria and oedema. The oedema affects the face on waking causing a characteristic puffy appearance; as the day wears on this disappears and oedema of feet and ankles develops: oliguria is common and there may be hypertension.

Prognosis: good: the disease usually clears up spontaneously: rarely, it may cause permanent kidney damage and eventually progress to renal failure. Second attacks are uncommon.

Pathogenesis. Pathologically there is increased cellularity of the glomeruli with larger deposits on the outer and smaller deposits on the inner surfaces of the basement membrane. The disease is the result of an immunological process but the exact pathogenesis is unclear. The deposits are immune complexes thought to be formed by combination of antistreptococcal antibody with either: (i) streptococcal antigens already on the basement membrane or (ii) streptococcal antigens circulating in the blood to be later deposited on the membrane after complex formation.

Laboratory diagnosis. Diagnosis is usually on clinical grounds: attempts should be made to confirm past or present streptococcal infection as described above for rheumatic fever.

Although less common than throat infection as a precipitating cause of the disease, streptococcal impetigo has caused some outbreaks of glomerulonephritis in the USA and in some tropical countries. It should be diagnosed by culture of a swab or sample of pus from the skin lesion.

Complement estimations. The level of C3 in serum is reduced: this has been interpreted as evidence of immune complex formation.

Use of antibiotics. Eradicate *Strep. pyogenes* if the organism is still present in the throat: penicillin is the drug of choice.

Diphtheria

Cause: Corynebacterium diphtheriae.

A severe disease: the inflamed fauces are covered with patches of sero-cellular exudate which forms a grey-white membrane. The severity is related to the extent of the membrane and the type of infecting strain of *C. diphtheriae*. Nasal diphtheria is often mild: laryngeal diphtheria is serious because of the risk of respiratory obstruction. *C. diphtheriae* produces a potent *exotoxin* which is *cardiotoxic* and fatalities from the disease are usually due to heart failure; it is also *neurotoxic* and can cause cranial and peripheral nerve palsies.

Incubation period: 2–5 days.

Source. Diphtheria is now very rare in developed countries due to effective active immunisation in childhood. The source of infection is the throat and (especially) the nose of symptomless carriers and of cases in which the disease is mild or inapparent. Convalescent carriage rarely lasts more than a few weeks.

Treatment. Antitoxin must be injected *without delay* — i.e. without waiting for bacteriological confirmation. Tracheostomy may be necessary to relieve respiratory obstruction. Penicillin or erythromycin should be given since antibiotics help to eliminate *C. diphtheriae* from the throat and, therefore, prevent further toxin production. Erythromycin is the preferred drug for the treatment of persistent carriers.

Candidosis

Oral thrush due to the yeast *Candida albicans* presents as white patches superimposed on red, raw mucous membrane which may involve the throat as well as the more common site of the mouth.

Source: endogenous; although candidosis may be precipitated by antibiotic treatment the patient is often debilitated by disease, e.g. malignancy and especially leukaemia.

Treatment: locally applied nystatin or amphotericin B.

Vincent's angina

An ulcerative tonsillitis which causes much tissue necrosis: often an extension of similar disease of gums and mouth (gingivostomatitis).

Source: endogenous. The causal organisms are a spirochaete (*Borrelia vincentii*) and Gram-negative anaerobic bacilli (*Fusobacterium species*) found in small numbers in the normal mouth. Overgrowth to produce disease is precipitated by dental caries or poor oral hygiene, nutritional deficiency, leucopaenia (e.g. in leukaemia) and viral infections (e.g. herpes simplex, infectious mononucleosis).

Treatment: penicillin and/or metronidazole.

Infectious mononucleosis

Exudative tonsillitis is often the presenting feature of this generalised viral infection due to Epstein-Barr virus — the so-called anginose form of infectious mononucleosis.

Source: oropharyngeal secretions — the kissing disease of young adults.

Treatment. No specific treatment. Antibiotics should be avoided: almost all patients given ampicillin develop a skin rash.

DIAGNOSIS OF THROAT AND PHARYNGEAL INFECTIONS

Isolation — or demonstration — of the causal bacterium.

Specimen: a well-taken throat swab. Illumination of the throat and depression of the tongue are essential. The swab should be gently rubbed over the affected area so that it collects a sample of any exudate present.

Gram-stained film: a mixed bacterial flora is always present and the only findings of value are the recognition of Vincent's organisms and yeasts.

Note. This is the only method of diagnosing Vincent's infection — the causal organisms cannot be isolated by routine culture methods.

Culture: the swab is inoculated onto a variety of media incubated at 37°C for 24–48 h:

i. Blood agar

ii. Crystal violet   — selective for *Strep. pyogenes*
   blood agar   especially if incubated anaerobically

iii. Sabouraud's medium   — selective for *C. albicans*

iv. Loeffler's serum slope ⎤ — for the isolation of *C. diphtheriae*
    Blood tellurite medium ⎦   not routine nowadays

2. INFECTIONS OF MIDDLE EAR AND SINUSES

Acute infection of the middle ear or sinuses is often due to secondary bacterial invasion following a viral infection of the respiratory tract: this may be a common cold or measles — of which otitis media is a frequent complication.

Acute infections

Otitis media: an upper respiratory infection involving the middle ear by extension of infection up the Eustachian tube. Predominantly a disease of children: the main symptom is *earache*.

On examination the eardrum is injected or red and the infection may progress to cause bulging with eventual rupture of the tympanic membrane and discharge of pus from the ear. Recurrent attacks are common.

Sinusitis: mild discomfort over the frontal or maxillary sinuses due to congestion is a frequent symptom in common colds: severe pain and tenderness with purulent nasal discharge, however, indicate bacterial infection and require treatment.

Causal bacteria: Haemophilus influenzae, Streptococcus pyogenes, Streptococcus pneumoniae.

Source: endogenous spread of organisms from the normal flora of the nasopharynx.

Diagnosis. In the majority of cases of sinusitis and in many of otitis media, specimens from the site of infection cannot be obtained. If the eardrum ruptures or if myringotomy (incision of the tympanic membrane to release pus in the middle ear) is performed, collect a swab of exudate; if drainage or lavage of the sinuses is carried out, material should be collected and cultured in the same way as a sample of pus on a range of suitable media.

Treatment: amoxycillin or ampicillin, alternatively erythromycin.

Chronic infections

Chronic suppurative otitis media. Characterised by suppuration in the middle ear distorted by chronic pathological changes. Periods of quiescence when the ear is relatively dry are followed by exacerbations when there is profuse discharge associated with pain. This is usually a long-standing disease which can recur at intervals throughout childhood and into adult life.

Chronic sinusitis. Painful sinuses with headache are prominent symptoms; often associated with nasal obstruction with mucoid or purulent nasal discharge.

Causal bacteria: are the same as those implicated in acute infections — i.e. infection is usually endogenous with bacteria from the normal upper respiratory flora. A variety of other organisms may be found, including *Staph. aureus* and a wide range of coliform bacilli. Pseudomonas and proteus are common in chronic ear discharges. The clinical significance of some of these organisms is uncertain.

Diagnosis. Swabs of pus from the ear; lavage specimens from the sinuses — such saline washings are always contaminated by nasal flora. Examine as specimens of pus.

Treatment of chronic infections. Antibiotics give disappointing results: whenever possible therapy should be guided by antibiotic sensitivities of isolated organisms but treatment may have to be on a 'best-guess' basis. Topical antimicrobials (e.g. neomycin or framycetin, polymyxin, bacitracin) are given in chronic otitis media since systemic drugs fail to penetrate to site of infection but there is little evidence that they are effective.

3. INFECTIONS OF TRACHEA AND BRONCHI

Laryngitis, tracheitis and bronchitis are usually associated with or follow a viral infection of the upper respiratory tract.

Laryngitis

Clinical features: hoarseness and loss of voice: in more severe form, *croup* (or acute laryngotracheobronchitis) with croaking cough and stridor. In children, most often associated with parainfluenza virus infection: occasionally due to a rare but important bacterial infection — acute epiglottitis.

Acute epiglottitis

Clinical features: severe croup syndrome in children (usually aged between 2 and 7 years) which may rapidly progress to respiratory obstruction and death. The epiglottis is inflamed and oedematous.

Causal bacterium: capsulated strains of *Haemophilus influenzae* (usually of type b).

Diagnosis: H. influenzae may be isolated from the epiglottis and from blood culture.

Treatment: parenteral amoxycillin or ampicillin: tracheostomy may be necessary.

Bronchitis

Clinical features: a feeling of tightness in the 'tubes', cough, initially dry and painful, later productive with expectoration of yellow-green sputum most marked in early morning specimens; variable degree of fever and of constitutional upset. Abnormal chest signs, e.g. rhonchi are found on auscultation.

Acute bronchitis in a patient with a healthy respiratory tract is often a trivial complication of a viral upper tract infection: the initial viral attack damages respiratory mucous membrane with paralysis of ciliary movement. Although viral acute bronchitis is usually mild and self-limiting, secondary bacterial infection often supervenes in more severe attacks especially in patients with chronic respiratory disease such as chronic bronchitis, asthma and bronchiectasis.

Chronic bronchitis. Acute exacerbations of chronic bronchitis are serious events in the course of a major killing disease: chronic bronchitis is not itself due to infection; aetiological factors include low socio-economic class, urban dwelling (atmospheric pollution) and tobacco consumption, especially cigarette addiction ('smoker's cough'). Exacerbations, however, are associated with bacterial infection. They commonly follow respiratory infections or a fall in atmospheric temperature with increase in humidity (together causing foggy weather): all these factors are often present concurrently in winter. During exacerbations both volume and purulence of sputum increase.

Pathological changes in chronic bronchitis are: (i) increase in the number of mucous-containing cells in the bronchi with consequent hypersecretion of mucus; (ii) inflammation, fibrosis, collapse, dilatation and cyst formation in bronchioles and alveoli. After exacerbations some changes may resolve but others do not, resulting in progressive, irreversible damage.

Causal bacteria and source:

1. *Haemophilus influenzae* (usually non-encapsulated strains) — the closely related organism — *H. para-influenzae* is sometimes isolated but is of doubtful pathogenicity.

2. *Streptococcus pneumoniae* (pneumococcus).

   H. influenzae and pneumococci are together present in the upper respiratory tract in health. The secondary bacterial invaders in bronchitis are therefore endogenous. Normal subjects have a sterile bronchial tree but in chronic bronchitis the bronchi become colonised, especially with *H. influenzae*, even when the disease is quiescent. During exacerbations the concentration of *H. influenzae* in respiratory secretions increases along with sputum purulence. Specific antibodies to *H. influenzae*, absent in non-smoking healthy adults, are present in the serum of two-thirds of chronic bronchitics. *H. influenzae* is now regarded as the prime pathogen in exacerbations of chronic bronchitis.

3. *Mycoplasma pneumoniae:* often unrecognised as an aetiological agent in bronchitis (and upper respiratory tract infections) which it causes much more often than it does pneumonia. Patients are usually school-age children and young adults. Source is exogenous and spread is by the respiratory route. Cases are usually sporadic but there may be family outbreaks and, occasionally, institutional epidemics. The frequency of infection in the community varies from year to year.

TREATMENT — ACUTE BRONCHITIS

1. In previously healthy subjects, acute bronchitis usually subsides in 2–5 days and does not require antibiotic therapy.

2. The majority of patients with troublesome, purulent, acute bronchitis are given antibiotics prescribed on an informed best-guess basis. The drugs must be active against both *H. influenzae* and pneumococci: fortunately both of these bacteria have predictable sensitivity patterns — although there are some β lactamase-producing *H. influenzae* which are resistant to ampicillin/amoxycillin and also some pneumococci which are resistant to tetracycline.

Laboratory examination of sputum is essential if the patient fails to respond to an apparently adequate course of treatment.

TREATMENT — CHRONIC BRONCHITIS

Treatment should start as early as possible and chronic bronchitics should have an emergency supply of antibiotics to take whenever a cold 'goes to the chest': this shortens the duration and reduces the severity of exacerbations but, unfortunately, does not prevent deterioration of respiratory function.

Short-term treatment (5–7 days)

1. *Ampicillin or amoxycillin:* may be bactericidal in action in adequate dosage. Amoxycillin is preferred because of better absorption in the presence of food and ability to penetrate mucoid sputum: ampicillin only attains bactericidal levels against *H. influenzae* when the sputum is purulent.
2. *Tetracyclines:* bacteriostatic action, uncertain sputum penetration. An advantage is the activity against *M. pneumoniae*.
3. *Cotrimoxazole:* bacteriostatic action: the ratio of sulphonamide to trimethoprim in the sputum is 4 to 1 because of poor sulphonamide penetration (compared to synergistic bactericidal ratio of 20 to 1 found in blood).
4. *Chloramphenicol:* a reserve drug because of bone-marrow toxicity. Although bacteriostatic it is very effective due to excellent penetration and high activity against *H. influenzae*.

Long-term treatment

Winter-long suppressive tetracycline treatment for advanced chronic bronchitis was claimed to reduce the number and severity of exacerbations. Fashionable 20 years ago, this continuous bacteriostatic regimen has been abandoned in favour of short-term courses of bactericidal treatment.

Vaccines

Viral: the only vaccine available is that against influenza. Since chronic bronchitics often suffer severe exacerbations after influenza and are at risk of developing secondary bacterial pneumonia, it is recommended that they be vaccinated annually. The vaccine contains inactivated virus of currently circulating A and B strains and achieves protection of the order of 60 per cent.

Bacterial: polyvalent pneumococcal polysaccharide vaccines have been developed recently. They may be of value in preventing pneumococcal pneumonia in this 'at risk' group.

Diagnosis
See 'Diagnosis of bacterial chest infections' on p. 176.

Cystic fibrosis
Due to improved management more infants and children with this disease, transmitted as an autosomal recessive trait, survive to adult life.

This inherited defect leads to production of abnormally viscid mucus which blocks tubular structures in many different organs: the most disabling obstructive changes affect the lungs and chronic respiratory infection is a major problem.

Causal bacteria:

1. *Staphylococcus aureus* initially, tending to be replaced by:
2. *Pseudomonas aeruginosa*

Treatment: appropriate antistaphylococcal or antipseudomonal drugs alone or in combination, the choice dependent on in vitro antibiotic sensitivity results. Long-term administration may be required.

Pertussis (whooping-cough)

Clinical features
An acute *tracheobronchitis* of childhood.

Onset is insidious — initially a catarrhal stage with common-cold symptoms which lasts about 2 weeks: followed by a stage of paroxysmal coughing (2 weeks): residual cough persisting for a month or more is a common sequel.

Paroxysmal cough is a diagnostic feature: it consists of repeated violent exhalations with a distressing, severe inspiratory whoop: there is expulsion of tenacious clear bronchial mucus: vomiting is common.

Fatality is low: but morbidity may be high: there is a significant risk of developing subsequent chronic chest disease, e.g. bronchiectasis. Most acute deaths are in infants during the first year — and especially in the first 6 months of life.

Causal bacterium: Bordetella pertussis (types 1, 3: 1, 2, 3 and 1, 2). (*Bord. parapertussis* causes mild whooping cough and is relatively rare in Britain).
(A similar syndrome may be caused by adenoviruses (types 1, 2 and 5) and *M. pneumoniae*).

Incubation period: 1–3 weeks: often about 10 days.

Source: patients — most infective during catarrhal stage, becoming non-infective at end of paroxysmal stage.

Spread: airborne via droplets.

Diagnosis

Isolation: isolation from infected clinical cases is not easy, and diagnosis is usually based on symptoms. Organisms are much less numerous after catarrhal stage (i.e. when typical symptoms develop) and in immunised patients.

Specimen:

(i) *Cough plate* — held in front of mouth during a paroxysm of coughing

(ii) *Pernasal swab* — to sample naso-pharyngeal secretions

Inoculate: Bordet-Gengou medium or charcoal blood agar.

Incubate: 3–5 days at 35–36°C.

Observe: moist 'mercury-drop' colonies.

Identify: by slide agglutination with specific antisera.

Serology. Rising titres of antibody in *paired* serum specimens may be diagnostic in older children (over 1 year). Detected by complement fixation, agglutination or immunofluorescence tests. These may give evidence of infection when the opportunity for bacterial isolation has been missed.

Treatment: antibiotics are only of value if given within the first 10 days of infection i.e. during the catarrhal stage when the diagnosis may not be suspected. If secondary pneumonia develops it should be treated appropriately (depending on the antibiotic sensitivity of the organisms isolated).

Administration of ampicillin or erythromycin to the patient reduces the duration of infectivity and these drugs are successful chemoprophylactics when given to close contacts (e.g. siblings).

Vaccination: this controversial topic is discussed on p. 338.

4. INFECTIONS OF THE LUNGS

Pneumonia

The most severe and life-threatening of respiratory infections: antibiotic therapy has transformed the prognosis and most cases can be successfully treated.

Clinical features

Onset: sometimes abrupt, sometimes insidious when due to the extension of a pre-existing respiratory infection.

Symptoms: fever, rigors, malaise: respiratory symptoms include shortness of breath, rapid shallow breathing, cough, sometimes pleural pain: sputum may be initially tenacious and rusty — later becoming purulent.

Clinical investigation: signs of consolidation of lungs i.e. dullness on percussion, reduced air entry, moist rales: there is usually polymorphonuclear leucocytosis.

Radiology: indicates the site and extent of the consolidation:

Pneumonia is of three main types:

1. *Lobar (or segmental) pneumonia:* in which the consolidation is limited to one lobe or segment of the lung: the main type of pneumonia seen in previously healthy people — now relatively uncommon
2. *Bronchopneumonia:* the consolidation is scattered throughout the lung fields although it is mainly concentrated at the bases; the most common form of pneumonia — seen principally in the elderly and in patients with debilitating or chronic respiratory disease such as chronic bronchitis, bronchiectasis.
3. *Primary atypical or 'virus' pneumonia:* patchy consolidation of the lungs with widespread opacities— often out of proportion to the mild degree of illness and the few clinical signs: usually a self-limiting disease

Legionnaire's disease. An extremely severe pneumonia first recognised in 1976 amongst members attending an American Legion Convention; basically a form of pneumonia.

Table 29.1 lists the main organisms associated with the different types of pneumonia.

Table 29.1   Causes of pneumonia

Pneumonia	Main causal organisms
Lobar	*Streptococcus pneumoniae*
Bronchopneumonia	*Streptococcus pneumoniae*
	Haemophilus influenzae
	(rarely *Staphycococcus aureus,* coliforms)
Primary atypical pneumonia	*Mycoplasma pneumoniae*
	Coxiella burneti
	Chlamydia psittaci
Legionnaire's disease	*Legionella pneumophila*

CAUSAL AGENTS

1. **Streptococcus pneumoniae** (pneumococcus): the main cause of pneumonia and responsible for over 80 per cent of cases of lobar and nearly half the cases of bronchopneumonia: the most common serotypes associated with pneumonia are types 1, 3 and 8.
   Source: human respiratory tract.

Spread:

a. *Exogenous droplet transmission of virulent strains* (incubation period 1–3 days)

b. *Endogenous* due to downward spread of pneumococci from the flora of the nasopharynx

Treatment. Recommended drug: penicillin. Other effective drugs: ampicillin/amoxycillin, cephalosporins, erythromycin, cotrimoxazole.

2. **Haemophilus influenzae:** often underestimated as a cause of pneumonia; in infants usually due to capsulated strains: in adults, often but not always, complicating chronic respiratory disease (*note*, in pneumonia following exacerbations of chronic bronchitis the pneumococcus is the commonest bacterial cause).
Treatment. Recommended drug: ampicillin/amoxycillin. Other effective drugs: tetracycline, cotrimoxazole.

3. **Staphylococcus aureus:** a relatively uncommon cause of bronchopneumonia: probably most often seen in hospital patients: sometimes after influenza the cause of severe secondary bacterial pneumonia: often fulminating and rapidly fatal.
Treatment. Recommended drug: cloxacillin (penicillin if the infecting strain is sensitive). Other effective drugs: fusidic acid, lincomycins, erythromycin, cephalosporins.

4. **Coliforms** e.g. *Escherichia coli, Proteus species, Klebsiella species, Pseudomonas species:* rare causes of bronchopneumonia: their isolation from sputum often indicates merely colonization of the respiratory tract, e.g. after a course of antibiotic, and must be interpreted cautiously: as a cause of pneumonia they are most often encountered in hospital patients especially in the immunocompromised (e.g. patients with leukaemia) or in those on support ventilation under intensive care.
Treatment. Recommended drugs: gentamicin, tobramycin, carbenicillin. Other effective drugs: await results of laboratory sensitivity tests.
Friedlander's bacillus: a klebsiella of uncertain taxonomy; has been described as a cause of a rare pneumonia with much tissue destruction.

5. **Mycoplasma pneumoniae** (formerly known as Eaton agent): the main cause of *primary atypical* or *'virus' pneumonia*: school-age children and young adults are the group most affected.
Incubation period: about 3 weeks.
Spread: respiratory (droplet) spread to involve individuals, families and sometimes institutions (e.g. schools, military camps): the prevalence peaks every 3–5 years.

Treatment. Recommended drug: tetracycline. Other effective drug: erythromycin.

6. **Coxiella burneti:** responsible for the acute febrile disease Q fever; up to half of the patients have pneumonia — usually a patchy consolidation.
   Treatment. Recommended drug: tetracycline.

7. **Chlamydia psittaci:** the causal agent of ornithosis and psittacosis in birds: may infect man via inhalation of dried bird droppings: produces an acute influenza-like illness with patchy pneumonia.
   Treatment. Recommended drug: tetracycline.

8. **Legionella pneumophila:** cause of Legionnaire's disease; now being increasingly recognised as a cause of pneumonia. Patients are typically middle-aged smokers often in poor general health.
   Symptoms: initially influenza-like, the illness progresses to a severe pneumonia sometimes with respiratory failure; mental confusion is a prominent feature.
   Source: environment e.g. soil and water.
   Spread: by contaminated aerosols, e.g. via dust, air-conditioning systems, etc. Person-to-person droplet spread has not been recorded.
   Treatment. Recommended drug: erythromycin. Other effective drug: rifampicin.

Viruses
Pneumonia in infants is predominantly due to respiratory syncytial virus: primary influenzal pneumonia in adults is a rare but exceedingly severe complication of influenza and is almost invariably fatal: however, viruses on their own are uncommon causes of pneumonia and antibiotic treatment is therefore indicated for all patients diagnosed as having pneumonia.

TREATMENT OF PNEUMONIA

Choice of antibiotic before laboratory results are available
This is governed by the clinician's experience, his personal preference and also knowledge of what the infecting agent is likely to be.

The penicillins are the drugs of first choice: in patients with lobar pneumonia — when the infecting organism is likely to be pneumococci, alone, prescribe penicillin; in bronchopneumonia, give ampicillin or amoxycillin since *H. influenzae* may also be involved.

Therapy should be changed if the patient fails to respond or if laboratory investigation indicates a more appropriate antibiotic: tetracycline or erythromycin are useful second-line drugs.

In *severe* pneumonia where the causal organism is unknown, a combination of gentamicin and erythromycin is recommended for the widest possible antimicrobial cover.

TUBERCULOSIS

Although tuberculosis is not considered in this chapter (see Chapter 38) in any long-standing chest infection, the **possibility of tuberculosis must always be considered.**

Aspiration pneumonia, lung abscess

Aspiration pneumonia follows inhalation of vomit or sometimes a foreign body in an unconscious patient: the causal organisms are commensals of the upper respiratory tract — principally *Strep. pneumoniae* but anaerobic bacteria notably *Fusobacterium species*, *Bacteroides melaninogenicus* and also microaerophilic streptococci, are involved in the majority of cases.

Lung abscess is nowadays rare: usually due to obstruction of a bronchus or bronchiole, e.g. by an inhaled foreign body or to suppuration developing within an area of pneumonic consolidation: the infecting organisms are similar to those listed above — namely *Strep. pneumoniae* and anaerobic and microaerophilic bacteria of the upper respiratory flora: specimens for laboratory investigation must be obtained by aspiration, e.g. by needle biopsy of the abscess.

Treatment. Recommended drugs: penicillin and metronidazole.

Empyema

Literally, pus in the pleural space and nowadays a rare complication of pneumonia: sometimes tuberculous: laboratory diagnosis requires aspiration and treatment depends on drainage and removal of the infected fluid with appropriate antibiotic therapy: usually due to *Strep. pneumoniae, Staph. aureus*, with upper respiratory anaerobes often implicated.

DIAGNOSIS OF BACTERIAL CHEST INFECTIONS

1. *Isolation:* of causal pathogen from sputum (or, less commonly, from the aspirate of a pleural effusion or lung abscess). Blood cultures should also be taken since the infecting bacterium may be present in the blood of up to one third of patients with pneumonia.
2. *Serological:* demonstration of specific antibody in patient's serum is only useful in cases of primary atypical pneumonia (see below).

Sputum examination and culture

Specimen: early morning sputum since this sample is likely to be the most purulent.

Collection: try to minimise salivary contamination — but the presence of some oro-pharyngeal flora is inevitable.

Transport: send to laboratory without delay: in transit, delicate organisms (e.g. *H. influenzae*) die, robust bacteria (e.g. coliforms) multiply and overgrow.

Laboratory examination
Macroscopic — note naked-eye appearance.
Microscopic:

1. *Gram film:* observe amount of pus, squamous epithelial cells (indicating buccal contamination) and, nature of the bacterial flora — if this is very mixed it is probably not significant but the presence of a predominant organism may allow an immediate provisional diagnosis (e.g. of a pneumococcal infection)

2. *Ziehl-Neelsen film:* examine for acid and alcohol fast bacilli; if present, a presumptive diagnosis of tuberculosis can be made. This examination is not always carried out nowadays because of the decline in the incidence of tuberculosis — but should always be done in immigrants or in areas where tuberculosis is still common (e.g. large cities) or if there is any clinical indication of tuberculosis

Culture: bacteria are not distributed evenly throughout sputum; select a purulent portion: alternatively homogenise the specimen by treatment with a liquifying agent: such treatment allows semiquantitative culture and may make evaluation of results easier.

Inoculate:

(i) Blood agar: observe for presence of a predominant organism and assess respiratory flora

(ii) Chocolate agar with bacitracin: selective for *H. influenzae*. Incubate (i) and (ii) at 37°C in air with 5–10 per cent carbon dioxide — this atmosphere is necessary for the primary isolation of some strains of *H. influenzae* and pneumococci

(iii) Blood agar incubated at 37°C anaerobically with 5–10 per cent carbon dioxide: this is not done routinely although pneumococci grow better under these conditions: indicated if infection with non-sporing anaerobes such as bacteroides is suspected e.g. in lung abscess or aspiration pneumonia

Assessment of culture results may be difficult because of contamination from the oropharyngeal flora. *Streptococcus viridans*, neisseriae, coagulase-negative staphylococci, commensal corynebacteria, etc. are normally regarded as upper respiratory tract commensals: small numbers of haemophilus and pneumococci in a mixed growth may be part of the normal flora; larger numbers, especially if other bacteria are scanty, are regarded as pathogens.

The oropharynx of patients who have received antibiotics often becomes colonised with coliform organisms and yeasts and, under such circumstances, undue significance should not be placed on their isolation.

Culture for Mycobacterium tuberculosis: not a routine examination nowadays but should be carried out if indicated clinically. Submit three samples collected on successive days for culture on Lowenstein-Jensen medium for 6 to 8 weeks.

Serological tests

Primary atypical pneumonia and Legionnaire's disease are difficult to diagnose by isolation of the causal organism. Suspected cases should be investigated serologically i.e. by detection of antibody in patient's serum: most often diagnosed by stationary high titres (although demonstration of a four-fold rising titre is better).

The tests used to diagnose these diseases are shown in Table 29.2.

Table 29.2   Serological tests in atypical pneumonias

Causal organism	Test used to detect antibody
Mycoplasma pneumoniae	Complement fixation
Coxiella burneti	Complement fixation
Chlamydia psittaci	Complement fixation
Legionella pneumophila	Indirect immuno-fluorescence

30

Diarrhoeal diseases

Most but not all cases of diarrhoea are due to infection: bacteria, viruses and sometimes protozoa cause diarrhoea but this chapter will deal principally with bacteria.

Host: the young are most susceptible; poor general health and nutrition also predispose to diarrhoeal disease.

Bacteria: factors such as size of infecting dose, enterotoxin production, ability to adhere to gastrointestinal epithelium and to invade the gut wall affect the ability of organisms to infect the gut.

Epidemiology: spread of diarrhoeal disease in a community largely depends on sanitation (i.e. adequate disposal of sewage), clean food and a safe water supply; personal hygiene (i.e. washing hands after defaecation) is a remarkably effective means of preventing faecal-oral spread.

Storage of food at room temperature must be avoided: this permits rapid bacterial multiplication and, with some bacteria, the formation of toxins.

INVESTIGATION OF DIARRHOEA

1. *History:* particularly with regard to

    (i) Clinical symptoms, duration, etc.
    (ii) Recent foreign travel.
    (iii) Food history including symptoms amongst other consumers of suspect food and probable incubation period.

2. *Specimens for laboratory examination:*

    (i) Faeces (rectal swab if none available).
    (ii) Vomit.
    (iii) The suspected food: strenuous efforts should be made to obtain samples: this is often difficult as it is usually eaten or discarded.
    (iv) Blood culture: in severe cases — especially the very young and the elderly.

Other investigations which are especially important in an outbreak include

1. Food-handling practices in the kitchen concerned.
2. Faecal samples from kitchen staff.

The main enteropathogenic bacteria with some characteristics of the diseases they cause are shown in Table 30.1.

Table 30.1   Causal organisms and some features of diarrhoeal diseases

Organism	Usual source	Common mode of spread
Salmonella species	Animal gut	Poultry; meat; milk
Staphylococcus aureus	Septic lesions on food handlers	Cooked meats, dairy products
Clostridium welchii	Animal gut	Stews; meat pies
Bacillus cereus	Environment (soil)	Rice
Campylobacter species	Animal gut	Poultry; meat; milk
Vibrio parahaemolyticus	Sea water	Shellfish
Escherichia coli	Human gut	Faecal-oral or via food, water, fomites
Shigella species	Human gut	Faecal-oral or via food, fomites
Vibrio cholerae	Human gut	Water and food
Clostridium difficile	Possibly human gut	Possibly faecal-oral
Yersinia enterocolitica *Yersinia pseudotuberculosis*	Animal gut	Faecal-oral or via food, water

SALMONELLA FOOD-POISONING

Probably the commonest cause of diarrhoea in Britain — and the incidence is still increasing.

Causal organisms: Salmonella species; there are more than 1500 serotypes but only a relatively small number are important or common causes of infection.

[Note. *Salmonella typhi, paratyphi A, B and C* cause enteric fever (Chapter 9) but *S. paratyphi B* sometimes also causes diarrhoea.]

Habitat: domestic animals, poultry.

CLINICAL FEATURES

Asymptomatic infections are frequent, so are cases with only mild gastrointestinal disturbance.

Main symptoms are the acute onset of abdominal pain and diarrhoea after an incubation period of 12–36 h: nausea is common and there may be fever and vomiting; dehydration is sometimes a problem especially in the young.

Septicaemia occasionally develops in severe cases.

PATHOGENESIS

Not well understood: site of infection may be the small or large intestine; some strains produce enterotoxins similar to those of toxigenic strains of *Esch. coli:* other salmonellae invade the mucosa of the small intestine like shigellae but the mechanism of infection with other types is unknown.

DIAGNOSIS

Isolation

Specimen: faeces.

Culture: on MacConkey's medium and desoxycholate citrate agar: Wilson and Blair's medium and Selenite F — subculture to MacConkey medium after 24 h incubation.

Observe: for pale (i.e. non-lactose fermenting) colonies.

Identify: initially by biochemical tests; then serologically to determine 'H' and 'O' antigens.

TREATMENT

Antibiotics are contraindicated except in septicaemic cases: they do not affect symptoms and may prolong convalescent carriage of the organism: they also contribute to emergence of antibiotic-resistant strains.

Treatment is rarely necessary: rehydration is occasionally required.

EPIDEMIOLOGY

Food derived from animals and poultry is the main source: this is usually meat, sometimes milk. Human cases and carriers may also play a role.

Food, especially meat and offal is often contaminated in the raw state and may subsequently be inadequately cooked: alternatively, salmonellae from raw meat may contaminate other cooked foods, e.g. via utensils, work-surfaces in which they are able to multiply.

Outbreaks of salmonella food-poisoning are common: they often involve communal catering, e.g. weddings, large dinners, etc., but may be a problem in hospitals especially in mental or geriatric units.

Serotypes. The most common is *Salmonella typhimurium*; other common salmonellae are *S. hadar*, *S. agona* and *S. heidelberg*.

CONTROL

Difficult: control is, in fact, clearly unsuccessful at present. It depends on:
1. Control of imported animal feedstuffs.
2. Good farming and abattoir practice: restricted prescribing of antibiotics especially to calves.
3. Rigorous hygiene in the kitchen, e.g. separation of cooked and raw foods, prompt and efficient refrigeration of cooked food, thorough thawing of both frozen poultry and meat before cooking.
4. Good personal hygiene in food handlers.
5. Exclusion of known human excretors from food handling.

STAPHYLOCOCCAL FOOD POISONING

The classic example of toxic food-poisoning — due to ingestion of food contaminated with the enterotoxin of *Staphylococcus aureus*: very rapid in onset because it is due to pre-formed toxin in food.

Causal organism: Staph. aureus strains which produce an enterotoxin: there are five antigenically distinct enterotoxins — A, B, C, D, E: enterotoxin-producing strains belong to phage group III.

CLINICAL FEATURES

Symptoms: acute onset of nausea and vomiting within a few hours of eating the contaminated food, sometimes followed by diarrhoea; self-limiting and rarely severe; dehydration is occasionally a problem in the uncommon severe case.

PATHOGENESIS

Pre-formed toxin: ingested in contaminated food has a local action on the gut mucosa. Since the toxin resists temperatures that kill *Staph. aureus,* food may contain toxin but no viable staphylococci.

DIAGNOSIS

Isolation

Specimens: the suspect food, vomit or faeces.

Culture: for *Staph. aureus* on ordinary media or mannitol salt agar.

Identify: by coagulase test; later phage typing to correlate identity of strains from food and patients.

Demonstration of enterotoxin

Specimens: food or vomit; culture-filtrate from an isolated strain of *Staph. aureus.*

Examine: by gel diffusion using specific antisera.

Note. Tests for toxin are available at Reference Laboratories.

TREATMENT

The disease is short and self-limiting so treatment is unnecessary.

EPIDEMIOLOGY

Source of infection: is usually a staphylococcal lesion on the skin especially fingers of a food handler.

Route of infection: ingestion: enterotoxin-producing strains of *Staph. aureus* multiply in the food liberating toxin: the toxin is relatively heat-stable and unless the food is afterwards thoroughly heated, retains activity.

Food: usually cooked food. Contamination most often takes place after initial cooking: cold cooked meat is often implicated; unless food is correctly stored at 4°C, staphylococci can multiply in the warm conditions of the kitchen with consequent toxin production. Other foods include dairy products, e.g. creams and custards.

Outbreak: the international nature of staphylococcal food-poisoning is vividly illustrated by a large outbreak in Japanese air travellers who became ill while *en route* from Tokyo to Denmark after eating ham contaminated by a food handler in Alaska.

CONTROL

Food hygiene: exclude handlers with septic lesions; prompt refrigeration of food after preparation.

CLOSTRIDIUM WELCHII FOOD-POISONING

A common form of food-poisoning due to contamination of food by anaerobic spore-bearing (and therefore heat-resistant) organisms.

Causal organism: Clostridium welchii: classically non-haemolytic strains with particularly heat-resistant spores but β-haemolytic strains with relatively heat-labile spores are being increasingly implicated.

CLINICAL FEATURES

Onset is acute — between 8 and 24 h after eating contaminated food: diarrhoea and abdominal pain are the predominant symptoms and vomiting is rare; the illness is self-limiting.

PATHOGENESIS

Heat-resistant spores of Cl. welchii survive 100°C for 30 min: during cooling after cooking, spores germinate into vegetative bacilli. These multiply rapidly if food is stored at room temperature (the temperature range for growth of *Cl. welchii* is 15–50°C) and if there are anaerobic conditions.

Following ingestion of food heavily contaminated with vegetative *Cl. welchii* sporulation takes place in the gut with liberation of a heat-labile enterotoxin which acts mainly on the small intestine.

DIAGNOSIS

As strains which cause food-poisoning also form part of the normal flora in 5–30 per cent of people, laboratory diagnosis of an individual case is not possible. However, there are more than 60 serotypes of *Cl. welchii* and isolation of the same serotype from most of the victims of an outbreak of food-poisoning and from suspect food if available is strong presumptive evidence that it is the cause.

Isolation

Specimens: faeces from as many as possible of the patients involved in the outbreak; samples of suspected food.

Culture: aminoglycoside blood agar anaerobically.

Observe: for typical colonies β-haemolytic or non-haemolytic.

Identify: by Nagler reaction: serotype by slide agglutination in a specialist reference laboratory for epidemiology.

TREATMENT

Sometimes rehydration: antibiotic therapy is unnecessary.

EPIDEMIOLOGY

Sources of infection: Cl. welchii is present in very large numbers as a commensal in the animal and human intestine; it is also ubiquitous in the environment.

Route of infection is the ingestion of food in which *Cl. welchii* has multiplied: because it is an anaerobic organism, cooked and reheated meat pies, stews and gravies provide a suitable environment for bacterial multiplication.

CONTROL

Food hygiene: especially adequate cooking of meat and meat products with prompt refrigeration if stored before consumption.

FOOD-POISONING DUE TO BACILLUS CEREUS

Typically associated with Chinese restaurants because of their frequent use of rice.

Causal organism: Bacillus cereus, an aerobic, spore-forming Gram-positive bacillus.

CLINICAL FEATURES

There are two distinct types of illness:

1. *Short incubation period:* 1–2 h, with nausea and vomiting often followed by diarrhoea; associated with bulk-prepared rice. This is by far the commoner type.
2. *Longer incubation period:* 6–16 h, sudden onset of abdominal pain and diarrhoea; associated with soups and sauces.

PATHOGENESIS

Unclear: possibly due to an enterotoxin.

DIAGNOSIS

Isolation

Specimens: suspected food, vomit, faeces.
Culture: on ordinary media.
Observe: for typical 'curled hair' colonies.

TREATMENT

The disease is self-limiting.

EPIDEMIOLOGY

Source: B. cereus and its spores are widespread in soil and cereals are usually contaminated with it.

Route of infection: some spores survive cooking; if storage is at a warm temperature there is germination into vegetative bacilli which multiply and produce toxin.

CONTROL

Food hygiene — correct storage of cooked food: reheating should be rapid.

CAMPYLOBACTER ENTEROCOLITIS

Recently improved techniques of faeces culture have led to this organism being recognized as a major cause of diarrhoea.

Causal organism: Campylobacter species. Campylobacters are small vibrio-like organisms: the nomenclature of individual species is confused; the most common in human infections appear to be due to the thermophilic species *C. coli* and *C. jejuni*.

CLINICAL FEATURES

Two clinical presentations:
1. *With prodromal symptoms:* of fever, headache, backache, limb pain, nausea and abdominal pain: after 24 h — sometimes longer — diarrhoea develops.
2. *Without prodromal symptoms:* acute onset of abdominal pain and diarrhoea.

Incubation period: 3 to 10 days.

Diarrhoea is often severe with blood and mucus in stools: up to 20 stools a day may be passed and faecal incontinence can be a problem; *abdominal pain* is a prominent feature of campylobacter infection.

Duration is often several days and relapses are common.

PATHOGENESIS

Infection typically involves the ileum but it is not restricted to the small intestine and there is often colitis also.

Histology: there is some suggestion from the results of gut biopsy that campylobacters are invasive and penetrate beyond the mucosa.

Septicaemia: follows in a small proportion of cases confirming the invasiveness of the causal organisms.

DIAGNOSIS

Isolation

Specimens; faeces.

Culture: on selective medium (contains vancomycin, polymyxin and nystatin) at 43°C.

Observe: for typical colonies.

Identify: by morphology — i.e. curved slender Gram-negative bacilli, characteristic darting motility, positive oxidase reaction, and other features.

TREATMENT

Usually self-limiting: erythromycin appears useful in relieving symptoms but indications for its use are not yet firmly established.

EPIDEMIOLOGY

Source of infection: farm animals — especially poultry — are probably the major source of human infection: milk has been incriminated; dogs and cats have also been reported as sources of campylobacters.

Route of infection: eating contaminated food; but there can be faecal-oral spread especially between children.

CONTROL

Food and personal hygiene: control of infection in animals is not practicable.

FOOD-POISONING DUE TO VIBRIO PARAHAEMOLYTICUS

Causal organism: Vibrio parahaemolyticus — a marine bacterium found in warm coastal waters and in fish and shellfish.

CLINICAL FEATURES

Symptoms: acute onset of vomiting and diarrhoea usually 8–24 h after eating raw sea food — especially imported shellfish.

Epidemiology: more prevalent in warmer months. Most common in Far East where raw fish is a delicacy: most cases in Britain are due to imported sea food but the organism exists in British sea waters.

ESCHERICHIA COLI DIARRHOEA

Although *Escherichia coli* is part of the normal commensal gut flora, paradoxically certain strains can be a cause of diarrhoea.

Esch. coli gastroenteritis can be divided into two categories depending on the kind of patients affected.

1. *Infantile gastroenteritis:* in infants under 2 years of age.
2. *Travellers diarrhoea:* in adults recently arrived in a foreign country.

Causal organism: diarrhoea-producing strains of *Eschi. coli* can be distinguished serologically from commensal strains by their 'O' or somatic antigens: some — but not all — such strains produce enterotoxin.

Esch. coli enterotoxins are of two types:

(i) *LT* — heat-labile toxin
(ii) *ST* — heat-stable toxin

The enterotoxin is plasmid-coded.

Certain other strains of *Esch. coli* possess *invasive properties* (like shigellae) whereas still other strains have K antigens and other surface structures which promote adherence to epithelial surfaces and which may be important in the pathogenesis of the diarrhoea.

Infantile gastroenteritis due to Esch. coli

Causal organism: specific serotypes of *Esch. coli* (e.g. 026, 055, 0111, 0126, etc.) which are known to be *enteropathogenic* strains.

CLINICAL FEATURES

Acute onset of diarrhoea after an incubation period which varies between 1 and 3 days: the diarrhoea may lead to dehydration and acid base imbalance.

Hypernatraemia is a particular problem because of the disproportionate loss of water relative to the sodium from the extracellular spaces.

Epidemiology: although infection is sporadic in the community, the main problem is outbreaks in nurseries.

Transmission: in nurseries and neonatal units, infection spreads directly from case to case and via environmental contamination (e.g. by communal toilets, shared equipment and the hands of the staff); in some neonates, their mother is the source of infection.

PATHOGENESIS

Esch. coli strains from some outbreaks produce enterotoxin(s) with a cholera-like action on the gut: however, strains from other outbreaks are non-toxigenic; the pathogenesis of the disease is complex and not yet fully understood.

DIAGNOSIS

Isolation
Specimens: faeces.
Culture: MacConkey's medium.
Observe: and pick several pink (lactose-fermenting) colonies for further tests.
Identify: serologically with polyvalent then individual specific O antisera.

TREATMENT

Rehydration with correction of fluid loss, and electrolyte and acid-base imbalance: antibiotic therapy is of doubtful value — it may be useful in severe cases.

CONTROL

Scrupulous hygiene in nurseries and neonatal units: examination of faeces of antenatal patients with diarrhoea and of all admissions to nurseries for presence of enteropathogenic serotypes.

Control of an outbreak: prompt isolation of cases and contacts; screening of staff to detect carriers of the epidemic strain; the unit should be closed to new admissions.

Travellers' diarrhoea (Turista)
Also known as 'Delhi belly', 'Montezuma's revenge', 'Tokyo two-step', etc.

CLINICAL FEATURES

Diarrhoea, abdominal pain and vomiting usually self-limiting and of a few days duration; occasionally protracted.
Route of infection: via contaminated food and drink; polluted water is the usual source of infection.

PATHOGENESIS

Enterotoxigenic strains of *Esch. coli* are responsible for many cases. The toxins have an action similar to that of *Vibrio cholerae* toxin.

DIAGNOSIS

Laboratory facilities are often not available at the tourist resorts: enterotoxin-producing *Esch. coli* have been demonstrated in some research studies.

CONTROL

Public health measures, e.g. clean water supply; caution in diet by travellers.

Other outbreaks of diarrhoea due to Esch. coli

Outbreaks in adults have been reported: for example, a large-scale outbreak of cheese-borne food poisoning has been attributed to *Esch. coli*; laboratory investigation of faeces for specific serotypes of *Esch. coli* is impractical in sporadic cases of diarrhoea.

BACILLARY DYSENTERY

Dysentery — 'the commonest of the unpreventable diseases' — is a worldwide problem and an important cause of death and morbidity in young children, especially in the Third World.

Causal organisms: Shigella species

1. *Sh. dysenteriae* (formerly *Sh. shiga*): 10 serotypes.
2. *Sh. flexneri:* six serotypes.
3. *Sh. sonnei:* one serotype (the main cause of dysentery in Britain).
4. *Sh. boydi:* 15 serotypes.

CLINICAL FEATURES

Incubation period: from 1 to 9 days.

Shiga dysentery

Due to *Sh. dysenteriae*: a severe, even life-threatening disease with fever, abdominal pain and diarrhoea; the stools contain blood, mucus and pus; the disease sometimes becomes septicaemic — indicating that, unlike other shigellae, *Sh. dysenteriae* has invasive properties.

Sh. dysenteriae produces a powerful neurological exotoxin but this probably does not play a role in Shiga dysentery: an enterotoxin and cytotoxin are also produced — their role is uncertain but they may be partly responsible for the organism's invasiveness.

Poor sanitary conditions provide the environment where this — the most severe form of dysentery — flourishes.

The elderly, the very young or inform are worst affected.

Geographic distribution: mainly found in India and other tropical countries (travellers returning with the disease have been misdiagnosed as suffering from ulcerative colitis).

Dysentery due to other shigellae

Dysentery — or *shigellosis* — due to the other members of the genus is generally a milder disease which varies from asymptomatic excretion to a prostrating attack of diarrhoea with abdominal pain and sometimes minimal fever: the stools may contain blood, mucus and pus — but blood is unusual in cases of Sonne dysentery:

Shigella sonnei: the usual cause of dysentery in Britain. Sonne dysentery, however, is worldwide in distribution: the main age group affected are young children, particularly those in nursery schools where epidemics are not uncommon; the disease is usually mild but in a few cases dehydration ensues requiring emergency treatment; epidemics of Sonne dysentery are also frequent in mental hospitals and the infection may be difficult to eradicate.

Shigella flexneri was formerly fairly common in Britain: it persisted as a major cause of dysentery in a few areas such as Glasgow long after it had disappeared from the country as a whole; improved housing conditions are thought to have been responsible for its virtual disappearance in Glasgow in recent years; still common overseas — mainly in tropical countries.

Shigella boydii is rare in Britain and the infection has almost always been acquired overseas: common in the Middle and Far East.

DIAGNOSIS

Isolation

Specimens: stools, rectal swabs.

Culture: MacConkey and desoxycholate citrate agar: Selenite F broth with subculture to MacConkey's medium after 24 h incubation.

Observe for pale (non-lactose fermenting) colonies but *note: Sh. sonnei* is a late lactose fermenter and may produce pale pinkish colonies.

Identify: by biochemical tests then serologically by antiserum to *Sh. sonnei* and, if necessary, polyvalent antisera to other *Shigella species.*

TREATMENT

Antibiotics are rarely necessary for Sonne dysentery: the more severe forms should be treated systemically depending on the sensitivity of the organism isolated.

EPIDEMIOLOGY

Reservoir of infection is the human gut: not only are symptomless infections common — with excretion of the organism in the faeces — but after an acute attack, a proportion of patients continue to excrete shigellae for some time, i.e. for weeks, sometimes months; patients with acute dysentery, however, are the most dangerous sources of infection doubtless due to the extremely large number of shigellae excreted during the acute phase of the disease.

Route of infection is faecal-oral either directly or via contaminated equipment and towels: contamination of lavatory seats is a particularly common source of infection in nursery schools and shigellae can remain viable for long periods of time in the cool, moist environment of the average nursery school toilet.

'Food, flies, fomites' are the classical means of spread of dysentery; contaminated water can also be a source of infection.

Dysentery waxes and wanes in incidence over long periods of time: after two decades of high incidence which started in 1950, the disease is now waning in Britain; this periodicity appears unrelated to improved sanitation or other public health measures.

CONTROL

Good sanitation with safe water, adequate sewage disposal: a high standard of personal hygiene is important but almost impossible to achieve in nursery schools.

CHOLERA

The world at present is experiencing the seventh pandemic of cholera: the six previous pandemics were during the nineteenth and early twentieth centuries. Glasgow Royal Infirmary is built on the site of mass cholera graves from the epidemic of 1849. The factors which determine the onset of epidemic spread are unknown.

Causal organism: Vibrio cholerae of which two biotypes are recognised.

1. *Vibrio cholerae* — the cause of classical cholera.
2. *Vibrio cholerae El Tor* — responsible for the present pandemic: causes a generally milder disease.

CLINICAL FEATURES

Acute onset — after an incubation period of from 6 h to 5 days — of abdominal pain and diarrhoea: the diarrhoea which is typically of exceptional severity progresses to the continuous passage of *'rice-*

water' stools; vomiting, dehydration, acidosis and collapse may follow; some cases, however, are much less severe with only mild diarrhoea.

PATHOGENESIS

Exotoxin: V. cholerae produces a potent protein exotoxin which is plasmid-coded: the toxin stimulates the activity of an enzyme — adenyl cyclase — which raises the concentration of cyclic AMP in cells causing an increase in the flow of water and electrolytes into the bowel lumen; the fluid lost has relatively high concentrations of bicarbonate and potassium; *V. cholerae* is not invasive and does not penetrate the gut mucous membrane.

DIAGNOSIS

Isolation
Specimen: faeces.
Culture: on selective medium (e.g. TCBS agar).
Observe: for typical colonies.
Identify: by slide agglutination with polyvalent antiserum.

TREATMENT

Correction of dehydration by the intravenous administration of fluid and electrolytes to restore the acid-base balance: mortality can be reduced from more than 50 per cent to nil with fluid replacement treatment.

Tetracycline given orally or intravenously may help to limit the duration of diarrhoea and reduce fluid loss.

EPIDEMIOLOGY

Reservoir of infection is man — the human gut.

Spread is faecal-oral usually via sewage contamination of the water supply but sometimes via food contaminated by foul water, flies or unclean hands.

Symptomless carriers are common in epidemics — for example, the case : carrier ratio with *V. cholerae El Tor* may reach 1:100: they form the main source of infection.

Pandemic: since 1960, cholera El Tor has gradually spread from the Far East to reach Mediterranean Europe: air travel probably increases the risk of importation of the disease into cholera-free areas.

CONTROL

Good sanitation with a clean water supply and adequate sewage disposal together with personal hygiene are effective methods of controlling the spread of cholera: unfortunately in many areas of the world these methods are simply not practicable; carriers and cases should also, if possible, be isolated.

VACCINATION

A vaccine containing heat-inactivated organisms is available: the protection conferred is limited (as is the immunity following natural infection).

ANTIBIOTIC-ASSOCIATED COLITIS

Diarrhoea is a common side-effect of antibiotic therapy: usually mild and self-limiting but with occasional severe cases in which the colonic

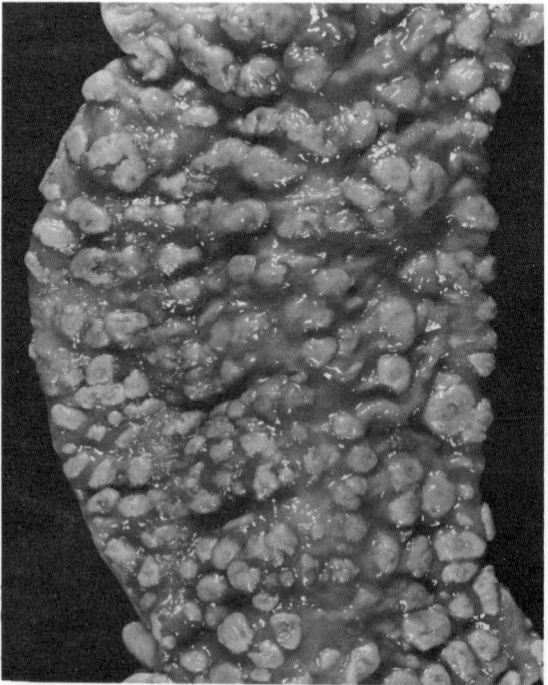

Fig. 30.1 Antibiotic-associated pseudomembranous enterocolitis. Segment of large bowel with typical pseudomembranous plaques. The plaques consist of fibrin and polymorphs and are the characteristic gross and microscopical lesions in the condition. (Photograph by Dr P.S. Macfarlane.)

mucosa has the characteristic appearance of 'pseudo-membranous' or antibiotic-associated colitis (Fig. 30.1).

Antibiotics: clindamycin, lincomycin and, to a lesser extent, ampicillin and tetracycline are most often implicated; many cases have followed administration of more than one antibiotic.

Causal organism: toxigenic strains of *Clostridium difficile*.

CLINICAL FEATURES

A life-threatening disease in which diarrhoea is the major symptom: the amount of diarrhoea varies but may be profuse (up to 20 stools per day): abdominal discomfort, fever and leucocytosis are often present.

Surgery: many cases follow abdominal operations.

PATHOGENESIS

Antibiotic therapy alters the balance of the normal flora: the changed environment may favour colonization, multiplication and toxin production by *Cl. difficile; Cl. difficile* is probably not part of the normal gut flora.

DIAGNOSIS

Clinically a high index of suspicion is necessary.

Proctosigmoidoscopy may show the characteristic membrane or isolated areas of white or yellow material adhering to the colonic mucosa: these areas should be biopsied for histological examination: the rectum is sometimes unaffected.

LABORATORY DIAGNOSIS

1. *Direct demonstration of toxin* in faeces by testing for cytotoxic effect on cells in tissue culture, e.g. HeLa or human embryonic lung cells: *Faecal filtrate* is prepared and tested for cytotoxicity and for neutralization of the toxic effect by specific antiserum.
2. *Isolation of Cl. difficile* from faeces with subsequent demonstration of toxigenicity.

   Note. Laboratory facilities for demonstration of *Cl. difficile* and its toxin are not generally available.

TREATMENT

Current antibiotic therapy must be stopped: oral vancomycin which is active against clostridia gives good results.

CONTROL

Antibiotics should be limited to cases in which there is a clear indication for their use: clindamycin and lincomycin should be avoided if at all possible.

Neonatal necrotizing enterocolitis

This condition, which is not antibiotic-associated, has been attributed to *Cl. butyricum.*

YERSINIOSIS

Probably an underdiagnosed disease but apparently not very common in Britain: most reported cases have been in Scandinavia and USA. *Causal organisms: Yersinia enterocolitica, Y. pseudotuberculosis.*

CLINICAL FEATURES

A septicaemic illness with abdominal pain, diarrhoea and fever: diarrhoea is more prominent in infections due to *Y. enterocolitica:* yersiniosis may mimic appendicitis and the disease in fact is associated with *acute terminal ileitis* and *mesenteric lymphadenitis.*

DIAGNOSIS

Isolation: from faeces or from blood cultures or mesenteric lymph nodes (if specimens of these are available).

TREATMENT

Tetracycline or aminoglycosides.

EPIDEMIOLOGY

Reservoir of infection: birds, wild and domestic animals.   *Spread:* via the faecal-oral route directly person-to-person or from animals; by contaminated food and water.

NON-BACTERIAL INFECTIVE DIARRHOEA

It must never be forgotten that diarrhoea may be due to infection with other non-bacterial agents. These are listed with some of their characteristics in Table 30.2.

Table 30.2 Non bacterial causes of diarrhoea.

Agent	Disease	Diagnosis	Treatment
Viruses Rotavirus	Acute non-bacterial gastroenteritis	Electron microscopy of stools	Symptomatic
caliciviruses	Winter vomiting disease	Electron microscopy of stools	Symptomatic
Protozoa *Entamoeba histolytica*	Amoebic dysentery	1. Microscopy of stools for trophozoites or cysts 2. Serology	Metronidazole
Giardia lamblia	Giardiasis	Microscopy of duodenal aspirate, stools for giardia trophozoites or cysts	Metronidazole, mepacrine

31

Botulism

A rare, severe disease due to ingestion of a bacterial toxin pre-formed in food: diarrhoea is not a symptom.

Causal organism: Clostridium botulinum, types A, B and E.

CLINICAL FEATURES

Incubation period: usually 12 to 36 h.

Symptoms: neurological — the toxin acts by reducing acetylcholine release at neuromuscular junctions; oculomotor, pharyngeal paralysis; vomiting, constipation, thirst, dryness of mouth, vertigo; sometimes difficulty in speaking.

Prognosis: often fatal; death is due to respiratory failure.

PATHOGENESIS

Cl. botulinum is a spore-forming anaerobe which forms an exceedingly potent exotoxin: found in soil, water, sludge; if spores contaminate food in conditions of anaerobiasis, germination follows and the vegetative bacilli multiply to produce toxin. Spore germination and toxin formation are inhibited by a low pH, e.g. they do not take place in acid fruits but can proceed in alkaline vegetables (e.g. string beans). The toxin is sensitive to heat and is destroyed by cooking.

DIAGNOSIS

1. *Specimen:* suspected food, patient's serum.
   Demonstrate toxin: by inoculation of mice.
   Observe: for paralysis and death.
   Identify: include in the test mice protected by antitoxins to types A, B and E toxins; protection by the appropriate antitoxin identifies the toxin present.

2. *Specimen:* suspected food.
Isolate: Cl. botulinum by pasteurisation of the food to destroy non-sporing bacteria followed by anaerobic culture at 35°C.
Identify: suspect colonies of *Cl. botulinum.*

TREATMENT

Antibiotics are of no value: treatment is supportive with artificial ventilation, etc.; antitoxin may help to neutralise absorbed toxin.

EPIDEMIOLOGY

Table 31.1 shows the habitat, geography and the kind of food associated with the three types of botulinum toxin that cause human disease.

Table 31.1 Medically important types of *Cl. botulinum*

Type	Habitat	Geography	Usual source of infection
A	Soil	USA USSR	Home-preserved vegetables, meat, fish
B	Soil	Europe USA	Meat, especially pork
E	Soil, sea water, sludge	Japan Canada Alaska	Raw or tinned fish

The disease is extremely rare in Britain but there was a small outbreak in Birmingham in 1978 due to tinned Alaskan salmon contaminated with type E toxin.

CONTROL

Amateur canning and preservation of food should be avoided. Commercial canning and pickling processes should be carefully controlled: a temperature that will kill the heat-resistant spores of *Cl. botulinum* is essential — e.g. 120°C for 20 min.

32

Enteric fever

Enteric fever includes typhoid fever and paratyphoid fever. Both are due to salmonellae which are markedly more invasive than the salmonellae which cause food-poisoning. Paratyphoid fever is generally a milder disease than typhoid fever.

TYPHOID FEVER

Causal organism: Salmonella typhi.

Clinical features

Incubation period: 14–21 days; sometimes longer.

Symptoms: a septicaemic, febrile illness with headache, toxaemia, dullness, apathy. Rose-coloured spots (which contain the infecting organism) are often seen as a sparse rash on the abdomen. Splenomegaly is sometimes present. There is generally some soft abdominal swelling and discomfort; leucopenia is common. Diarrhoea is a late symptom usually in the third week of illness.

Duration: untreated, about 4 weeks: symptoms clear up in around 3–4 days with antibiotic therapy.

Complications: intestinal perforation and haemorrhage are the most serious; rarely, periostitis, myocarditis, pneumonia.

Relapse: is common, with recrudescence of symptoms about a week after the end of the primary illness.

Carriers. Approximately 3 per cent of patients with typhoid fever become chronic carriers due to persistent infection of the gall bladder; this results in S. *typhi* being discharged into the gut and so into the faeces. Carriage is often associated with cholecystitis and gall stones.

PARATYPHOID FEVER

Causal organism: S. paratyphi A, B and C; S. paratyphi B is common in Britain; S. *paratyphi A and C* are virtually confined to tropical countries.

Clinically a milder febrile illness of shorter duration and incubation period; transient diarrhoea and symptomless infections are common.

Carriers: the percentage of cases who become carriers is lower than in typhoid fever.

The text of the rest of this chapter applies to both typhoid and paratyphoid fevers.

PATHOGENESIS

The way in which *S. typhi* and *S. paratyphi* invade the body to cause disease is shown below.

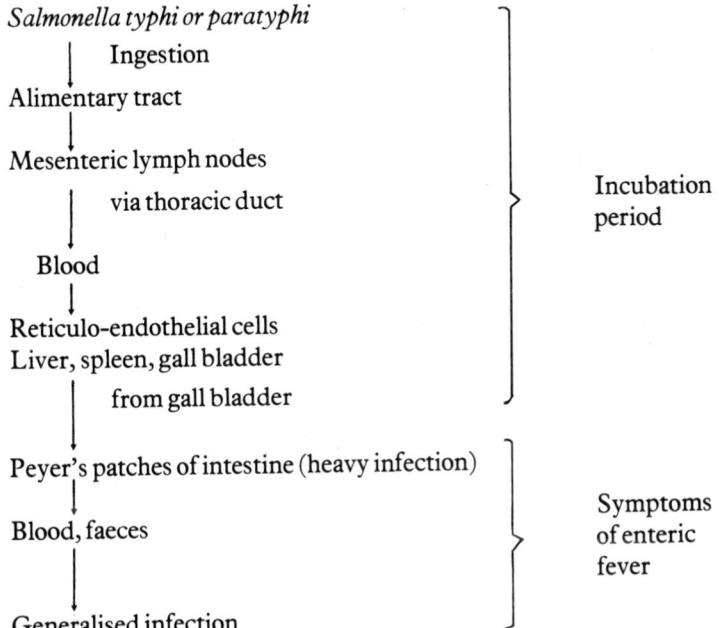

DIAGNOSIS

Isolation

Specimens: faeces, blood, and urine. Blood culture is positive in over 80 per cent of patients in the first week of illness.

Faeces culture:
a. Solid selective media:
   (i) MacConkey's medium
       Observe: for non-lactose-fermenting (pale) colonies
   (ii) Desoxycholate citrate agar

Observe: for non-lactose-fermenting (pale) colonies
(iii) Wilson and Blair bismuth sulphite agar
Observe: for shiny black metallic colonies

b. Fluid enrichment media:
   (i) Tetrathionate broth
   (ii) Selenite F
   Subculture: after 24 h onto MacConkey's medium
   Observe: for non-lactose-fermenting pale colonies

Identification:

1. *Biochemical reactions* (API test)
   Note S. typhi, unlike other salmonellae (including *S. paratyphi*) produces no gas on fermentation of sugars.
2. *Serological:* preliminary identification with salmonella polyvalent 'H' and 'O' antisera. Final identification: send to Reference Typing Laboratory.

Bacteriophage typing: to distinguish between strains of *S. typhi* (and also of *S. paratyphi B*); useful for epidemiological investigation into the source of outbreaks.

Blood culture and urine culture: subculture blood cultures and culture urine directly on MacConkey's medium with subsequent identification as described above.

Serology
Widal test: agglutination test for antibodies to flagellar H antigens (using formalized bacteria) and somatic O antigens (using boiled bacteria) of *S. typhi, S. paratyphi A* and *B.*

Significant titre:

Titre: a titre of 80 to both H and O antigens suggests active infection in an unvaccinated patient living in an areas where enteric fevers are not endemic.

Vaccinated patients: a considerable proportion of the population has been vaccinated against typhoid and paratyphoid fever: in them, diagnosis of infection by Widal test is unreliable. H antibodies persist longer than O antibodies.

Vi antibody: antibody to the Vi somatic antigen is usually produced during typhoid fever: not normally tested: occasionally Vi antigen masks O antigen and Vi antibody is produced without production of O antibodies.

TREATMENT

Acute disease

1. *Chloramphenicol:* effective in clearing up symptoms rapidly; does not prevent relapse or carriage.
2. *Cotrimoxazole:* gave promising results in early trials; probably as good as chloramphenicol for rapid control of symptoms in acute phase.
3. *Ampicillin:* good in vitro activity is not matched by in vivo therapeutic effect.

Carriers

It is notoriously difficult to eradicate *S. typhi* from the gall bladder. Antibiotic therapy is effective in curing some carriers but, in a proportion, the infection persists and they become long-term, even permanent, carriers.

1. *Ampicillin:* probably the most successful drug; treatment must be in large doses and prolonged.
2. *Chloramphenicol:* less successful in carriers than ampicillin; long courses of treatment in any event should be avoided because of risk of erythropoietic damage.
3. *Cholecystectomy:* cures carriage but the operation carries a small mortality rate and is only resorted to under special circumstances.

EPIDEMIOLOGY

Habitat: the human gut.

Source of infection. Chronic carriers who excrete the organism; excretion in faeces — less commonly in the urine — continues for about two months after the acute illness in most patients: after both typhoid and paratyphoid fevers a proportion of patients become permanent carriers.

Route of infection. ingestion of water or food contaminated by sewage or via the hands of a carrier; direct case-to-case spread is very rare.

Infecting dose. Small numbers of *S. typhi* can cause typhoid fever — hence why water-borne infection is common despite the great dilution of organisms; large doses are required to infect in paratyphoid fever.

Sporadic cases are not uncommon in Britain: about three-quarters have acquired their infection abroad.

Outbreaks. Numerous — usually explosive — outbreaks are on record, several involving large numbers of people. There are two main types.

1. *Water-borne:* in which sewage containing organisms from a carrier, pollutes drinking water, e.g. the outbreaks in Croydon 1937 and in Zermatt in 1963.
2. *Food-borne:* in which food becomes contaminated via polluted water or via the hands of carriers. 'Typhoid Mary', possibly the most famous carrier, worked as a cook in the USA and caused numerous outbreaks there in the early years of this century. Typhoid and paratyphoid bacilli multiply readily in most types of food.

Tinned food may become contaminated during canning, e.g. the large outbreak in Aberdeen in 1964 was due to a tin of corned beef which had been cooled in sewage-contaminated water; bacteria entered the can through tiny holes in the metal casing.

Shellfish often grow in estuaries where the water may be polluted by sewage: if eaten uncooked, they may cause infection; in the past a common source of typhoid, but not paratyphoid, fever.

Milk or cream products contaminated through handling by carriers have caused many outbreaks of both typhoid and paratyphoid fever: artificial cream is particularly associated with *S. paratyphi B* infection.

Other foods such as meat products, dried or frozen eggs, dried coconut, have been responsible for infection as a result of contamination by handlers who were carriers.

Animals: S. paratyphi B, unlike *S. typhi,* occasionally infects cattle; this has caused some outbreaks but much less commonly than infection from human sources.

CONTROL

Public health. The most effective way of controlling typhoid and paratyphoid fevers is provision of a clean water supply, adequate arrangements for sewage disposal and supervision of food processing and handling.

Carriers. If refractory to treatment, they must not be employed in food preparation. If instructed in personal hygiene, are rarely a danger to family and close contacts.

Vaccine TAB. (see Chapter 47).

33

Urinary tract infections

Urinary tract infections remain a major clinical problem 40 years after the introduction of antibiotics: many consultations in general practice are because of urinary infections.

Urinary infection is defined as bacteriuria, i.e. the multiplication of bacteria in urine within the renal tract: a concentration of 100 000 organisms per ml is regarded as *significant bacteriuria*.

Pyuria is the presence of pus cells (polymorphs) in the urine: it usually — but not always — accompanies bacteriuria:

Infections of the urinary tract may involve:

1. **Bladder**
   Cystitis.
2. **Kidney**
   pelvis — *pyelitis*
   parenchyma — *pyelonephritis*
   Note. It is impossible to distinguish between these two conditions and is probably better to refer to both as pyelonephritis. Pyelonephritis may become a chronic infection.
3. **Urethra**
   Urethritis (see Chapter 40).

CLINICAL FEATURES

1. Dysuria, frequency, urgency, suprapubic pain, sometimes haematuria are the classical symptoms of cystitis.
2. Loin pain and tenderness, rigors and fever are signs of pyelonephritis.
3. Chronic pyelonephritis causes general ill health and malaise with nocturia.

Women. Urinary infection is predominantly a disease of women (sex ratio 10:1).

Structural abnormalities of the renal tract, e.g. congenital neurogenic bladder, prostatic enlargement, cause *obstruction* and *stasis* in the tract and greatly increase the risk of infection.

205

Symptoms of infection without bacteriuria are surprisingly common: probably only half the women who consult their doctors because of symptoms are actually infected; the cause is unclear and the condition is often referred to as the 'urethral syndrome'.

· *Symptomless urinary infection,* or 'covert bacteriuria', is, conversely, also not uncommon: it can be detected in 5 per cent of adult women, 1 per cent of girls and 0.3 per cent of boys (in whom it is often associated with abnormality of the renal tract).

Screening to detect, symptomless or covert bacteriuria has been advocated in schoolchildren and pregnant women.

CAUSAL ORGANISMS

Escherichia coli: the cause of 60–90 per cent of urinary infections.

Certain serotypes of Esch. coli are particularly common in urinary infection (e.g. 02, 04, 06, 07, 08, 018, 075, etc.): this is probably because they are often present in the colon rather than because of inherently high pathogenicity for the urinary tract; bacterial strains which are rich in capsular (K) antigen are reputed to be more invasive than others.

Staphylococcus saprophyticus: an important cause of infection related to sexual activity in women under 25 years old: detected in 30 per cent of such infections; — surprisingly — seldom isolated from faeces and the anogenital region of young women.

Proteus mirabilis: responsible for 10 per cent of infections.

Klebsiella species: often multiply-antibiotic resistant.

Streptococcus faecalis: often found accompanying infection with coliforms.

Pseudomonas aeruginosa ⎤ especially after catheterisation
Staphylococcus aureus ⎦ or instrumentation

Mycobacterium tuberculosis: renal tuberculosis is described in Chapter 38.

Acute uncomplicated urinary infection is usually due to one type of organism.

Chronic infection is often associated with more than one type of organism.

SOURCE, ROUTE AND FACTORS INFLUENCING URINARY INFECTION

Source: the reservoir of urinary pathogens is the flora of the colon.

Route of infection is ascending via the urethra from the perineum.

Neonates: haematogenous spread may play a role in renal infection in the newborn.

Female preponderance is probably due to the shortness of the female urethra: turbulence of urinary flow during micturition may result in bacteria entering the bladder.

Sterility of urine is maintained by 'flushing' (i.e. from the emptying of the bladder and constant in-flow of newly formed urine) and local defence mechanisms in the bladder wall.

Sexual intercourse is correlated with infection in young women — possibly due to retrograde 'milking' of the urethra during coitus ('honeymoon cystitis').

Incompetence of the vesico-ureteric valve: due to congenital abnormality or inflammation of the bladder wall causes reflux of urine into the kidney pelvis during micturition; this may lead to pyelitis and pyelonephritis.

DIAGNOSIS

Specimen: a mid-stream specimen or urine collected to avoid contamination from perineum or vagina: in babies, use a strategically placed self-adhesive plastic bag but suprapubic needle aspiration of the full bladder may be necessary.

Transport: bacteria multiply in urine so specimens must be submitted to the laboratory within 2–4 h of collection; if this is not possible, do one of the following:

1. Refrigerate the specimen at 4°C.
2. Use container with boric acid, a bacteriostatic preservative, to give a final concentration in urine of 1.8 per cent.
3. Use a dip slide coated on both sides with culture medium and inoculated by dipping into the urine: can be read even after several day's delay.

Direct examination:

Wet film: for the presence of pus cells and bacteria: erythrocytes and casts should be noted.

Gram film: not often required: sometimes useful for preliminary identification of bacteria.

Culture: semi-quantitative culture on media such as CLED or MacConkey agar (to prevent swarming of proteus).

Observe and count the number of colonies obtained (see Fig. 33.1 and 2).

(i) *More than 100 000 bacteria per ml:* evidence of urinary infection; carry out sensitivity tests with appropriate antibiotics

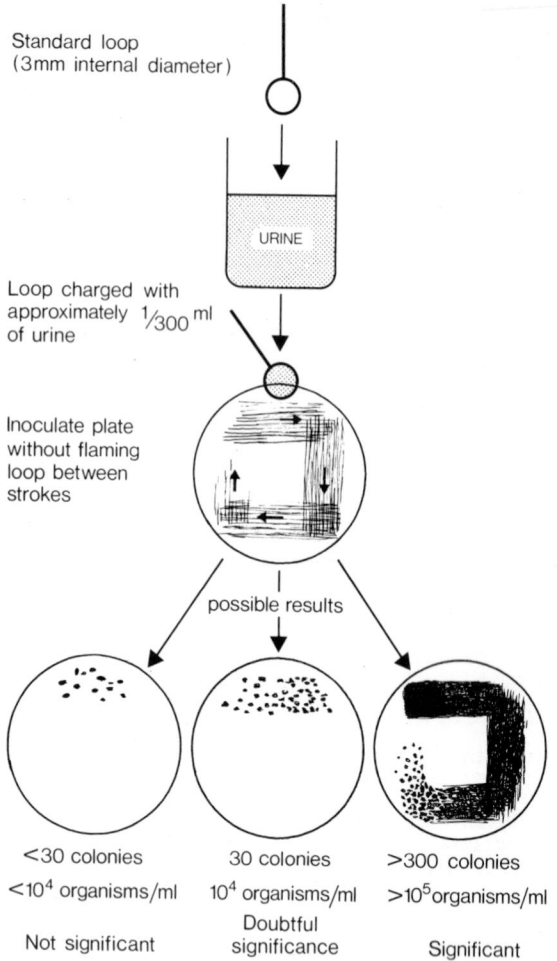

Standard loop
(3mm internal diameter)

URINE

Loop charged with
approximately $\frac{1}{300}$ ml
of urine

Inoculate plate
without flaming
loop between
strokes

possible results

<30 colonies	30 colonies	>300 colonies
$<10^4$ organisms/ml	10^4 organisms/ml	$>10^5$ organisms/ml
	Doubtful	
Not significant	significance	Significant

Fig. 33.1 Diagram illustrating the semi-quantitative culture of urine specimens.

(ii) *Between 10 000 and 100 000 bacteria per ml:* significance doubtful;
     further specimens should be obtained
(iii) *Less than 10 000 bacteria per ml:* regard as contaminants (unless
     the patient is on treatment for a known urinary infection)

Note. If there is pyuria without bacteriuria *renal tuberculosis* should be
suspected and three entire early morning specimens examined for
Mycobacterium tuberculosis.

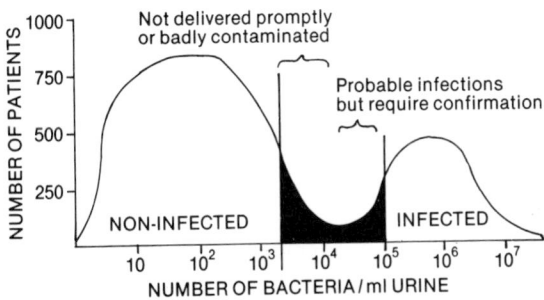

.**Fig. 33.2** Significance of urinary bacterial counts (Courtesy of Professor W. Brumfitt).

TREATMENT

Several suitable oral antibiotics are available: all are excreted in the urine in high concentration:

sulphonamide	nitrofurantoin
trimethoprim	nalidixic acid
cotrimoxazole	cephalexin or cephradine
ampicillin	

Chronic pyelonephritis: parenteral therapy with an aminoglycoside (e.g. gentamicin) may be required.

Cure rate of the acute infection is around 90 per cent but 50 per cent of the infections recur within a year.

Relapse. Recurrence within one month of stopping treatment is usually due to relapse — i.e. recrudescence of infection with the same organism: evidence of therapy failure.

Reinfection. Later recurrence is often a result of reinfection with a different organism.

Recurrent infections may indicate underlying abnormality of the urinary tract and require further investigation — e.g. urogram; management often requires long-term suppressive treatment with low doses of cotrimoxazole or nitrofurantoin.

In adults: bacteriuria — with or without symptoms — is nevertheless a relatively benign condition and permanent or progressive renal damage is rare.

In young children: under 5 years of age, the prognosis is much worse: bacteriuria with vesico-ureteric reflux often results in progressive renal damage with scar formation and impairment of kidney growth; some of these children go on to develop *chronic pyelonephritis* — which accounts for 20 per cent of end-stage renal failure.

SURGICAL URINARY INFECTION

A particular problem in urological and gynaecological wards usually following catheterization and instrumentation.

Catheterization: the risk of infection after a single catheterization — even if carefully carried out — is 5 per cent: almost all patients with an indwelling catheter develop infection.

Source of infection

1. *Endogenous:* from contamination of the patient's urethra or perineum by bacteria from the colonic flora.
2. *Exogenous:* due to cross-infection with bacteria from the infected urinary tract of another patient: transmission by instruments (e.g. cystoscopes, catheters) or hands of doctors and nurses.

Prevention

Careful attention to aseptic technique: use of disposable plastic catheters; introduction of antiseptics into the urethra before instrumentation.

Indwelling catheters should be attached to a closed drainage system to prevent retrograde bacterial spread into the bladder from the collection bag: prophylactic antibiotic therapy may be given to cover the immediate post-operative period.

34

Meningitis

Bacterial meningitis (classically "pyogenic" or polymorphonuclear) is a much more severe disease than viral meningitis (classically "aseptic" or lymphocytic): despite antibiotic therapy bacterial meningitis remains a serious cause of morbidity and mortality (Fig. 34.1) and is a bacteriological emergency requiring prompt diagnosis and treatment.

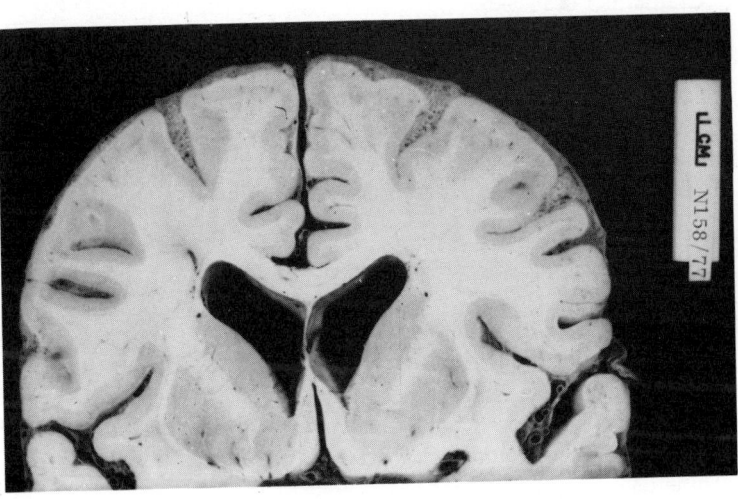

Fig. 34.1 Bacterial meningitis. Section of brain showing a thick layer of purulent exudate surrounding the brain and extending into the sulci. (Photograph by Professor J. Hume Adams.)

CLINICAL FEATURES

Symptoms: headache with malaise and fever; the onset is often abrupt: vomiting, photophobia and convulsions are not uncommon and patients may show irritability, apathy or drowsiness progressing to unconsciousness.

Signs of meningeal irritation — neck stiffness and pain on extending legs and thighs (Kernig's sign).

211

'*Meningism*' (apparent signs of meningeal irritation without meningitis) may be a feature of other types of severe infections: the differential diagnosis of bacterial meningitis has to be made from this and from subarachnoid haemorrhage.

Neonatal meningitis: the characteristic clinical features of meningitis are usually absent — the only presenting features being that the baby is obviously unwell with failure to feed and often vomiting. The condition is eight times more common in premature than in full-term babies.

Sequelae

Unfortunately, despite the advent of antibiotic therapy, neurological sequelae are not uncommon in survivors: the main types of sequelae are:

1. Encephalopathy — a state of altered consciousness (e.g. drowsiness) and convulsions.
2. Cranial nerve palsies.
3. Cerebral infarction
4. Obstructive hydrocephalus either external due to blockage of CSF drainage through arachnoid granulations or internal involving the ventricles.
5. Subdural effusion of sterile or infected fluid.

Age: meningitis is largely a disease of childhood.

CAUSAL ORGANISMS

1. Neisseria meningitidis (meningococcus)

The main cause of meningitis in Britain: affects all ages but most common in children and young adults. Of the eight serogroups of meningococci, groups A, B and C are responsible for most disease, the other groups being commoner in carriers than cases. Meningococcal meningitis is endemic and sometimes epidemic.

Incubation period: short, around 3 days.

Source: the reservoir is the human nasopharynx.

Spread: via infected respiratory secretions (i.e. 'droplet spread') from carriers and cases. Carriage rate in normal populations is about 5 to 10 per cent: this may rise to 50 per cent or more before and during epidemics in closed communities, e.g. institutions, military camps.

Route of infection: from nasopharynx via blood stream to meninges. During the *meningococcaemia* a characteristic *petechial rash,* rare in other types of meningitis, is a common finding. Possession of specific antibody prevents the meningococcaemia and meningitis.

Control: avoid overcrowding in living and working conditions.

2. Haemophilus influenzae
Most common in infants and pre-school children aged 3 months to 3 years. The causal strains are capsulated and serologically almost always of Pittman type b. A nasopharyngitis precedes spread to the blood stream followed by involvement of the meninges.

3. Streptococcus pneumoniae (pneumococcus)
Although also common in children this is the usual cause of meningitis in the middle-aged and elderly, often in patients who are in poor general health; this type of meningitis is often the sequel to pneumococcal infection of the lungs, sinuses or middle ear.

4. Mycobacterium tuberculosis
Seen in people of all ages, but most common in children 2 to 5 years old. The result of dissemination of infection from a primary focus usually in the lung.

Mortality rate
The overall case fatility rate in bacterial meningitis is around 7 per cent. Pneumococcal and tuberculous meningitis have the highest mortality, meningococcal meningitis the lowest.

Rare causal organisms:
1. *Listeria monocytogenes:* mainly seen in infants, the elderly and in compromised patients.
2. *Cryptococcus neoformans:* meningitis due to this fungus is rarely seen in previously normal people but is found in compromised patients, e.g. those with leukaemia or lymphoma.
3. *Leptospira interrogans.*

Neonatal meningitis: is mainly due to Gram-negative bacilli such as *Escherichia coli, Klebsiella species* and *Proteus species:* β-haemolytic streptococci of Lancefield group B are another important cause and are usually acquired from the mother's vagina. The baby is at greatest risk if there has been premature rupture of the membranes during labour or if the mother has a perinatal genital tract infection.

PATHOGENESIS
Spread to the CNS is by one of two routes:
1. *Haematogenous* via blood stream to the subarachnoid space.
2. *Direct* through the skull from infected foci in such sites as the middle ear, paranasal sinuses and nose via the cribriform plate.

Controversy exists as to which is the more important route: probably haematogenous but there may be direct spread when there is an adjacent focus of infection or following skull fracture.

DIAGNOSIS

Laboratory diagnosis of bacterial meningitis depends on examination of the CSF. Table 34.1 lists the results of CSF examination in meningitis compared to that in the two most important diseases in the differential diagnosis i.e. subarachnoid haemorrhage and cerebral abscess.

Red blood cells may be present in a normal CSF due to accidental damage to a blood vessel during lumbar puncture (universally known as a 'bloody tap') — in such cases the supernatant fluid after centrifugation is clear whereas in subarachnoid haemorrhage it is stained yellow to orange (xanthochromia).

Isolation:

1. *Specimen:* CSF obtained by lumbar puncture.
Examine:
a. In a counting chamber for nucleated cells and erythrocytes.
b. Gram film of centrifuged deposit for bacteria and cells.
c. If indicated, Leishman film of centrifuged deposit to differentiate polymorphonuclear leucocytes from lymphocytes.
d. If indicated, Ziehl-Neelsen film of centrifuged deposit for tubercle bacilli.

Culture:
(i) Centrifuged deposit onto blood-agar, chocolate agar and into glucose broth and cooked meat broth: incubate plates in air plus 5 per cent CO_2.
(ii) If indicated, Lowenstein-Jensen medium for culture for tubercle bacilli.

2. *Specimen:* blood culture; often positive in meningitis due to *N. meningitidis, H. influenzae* and *Strep. pneumoniae.*

Countercurrent immunoelectrophoresis: the CSF is tested for the presence of bacterial antigen by running it in an agar gel against specific antibacterial antisera. Of value when meningitis has been partially treated and no infecting organisms can be seen or cultured.

ANTIBIOTIC TREATMENT

Initial treatment should be parenteral by the intramuscular or intravenous route: later on antimicrobial drugs can be administered orally. Opinions differ on the need for intrathecal drug administration.

Table 34.1 Findings in cerebro spinal fluid

	Causal microorganisms	Appearance	Cells per mm³	Microbiology	Protein	Glucose
Normal	—	Clear, colourless	0-5 lymphocytes	Sterile	150 — 450 mg per litre	2.8 — 3.9 mmol per litre
Bacterial meningitis	*Neisseria meningitidis, Haemophilus influenzae, Streptococcus pneumoniae*	Turbid	500 — 20 000 mainly polymorphs, few lymphocytes	Bacteria in Gram-stained deposit. Growth on culture	Markedly raised	Reduced or absent
Viral (aseptic) meningitis	Enteroviruses, Mumps virus	Clear or slightly turbid	10 — 2000 mainly lymphocytes	Viruses rarely isolated from CSF Diagnosis by stool culture (entero-viruses) or serology (mumps)	Normal or slightly raised	Normal
Tuberculous meningitis	*Myobacterium tuberculosis*	Clear or slightly turbid	10 — 1000 mainly lymphocytes, polymorphs in early stages	AAFB in Z.N. stained deposit — often scanty. Growth on L.J. culture	Moderately raised	Usually reduced
Cerebral abscess	*Streptococcus milleri, Bacteroides species, Staphylococcus aureus, Proteus species*	Clear or slightly turbid	0 — 1000 mainly polymorphs, some lymphocytes	Organisms often not present in CSF	Normal or raised	Normal
Subarachnoid haemorrhage	—	Turbid, often blood-stained. Supernatant yellow-orange	Large numbers of red blood cells	Sterile	Markedly raised	Normal

1. *Meningococcal meningitis*

Penicillin is the drug of choice: sulphonamide penetrates the CSF well and was formerly widely used but its effectiveness has diminished due to the increased frequency of sulphonamide-resistant strains — in the UK at least 10 per cent of strains are fully resistant and a larger proportion are partially resistant.

Chemoprophylaxis is indicated for close contacts of a patient to eradicate the organism from the nasopharynx. Use sulphonamide (if the strain is sensitive) or rifampicin or minocycline (a tetracycline). Note that penicillin is *not* effective in this situation.

2. *Haemophilus influenzae meningitis*

Chloramphenicol penetrates readily to the CSF and is the drug of choice. Ampicillin is also effective — the appropriate alternative drug. Do not give these drugs in combination.

3. *Pneumococcal meningitis*

Penicillin is most widely used; the alternative is cephaloridine.

4. *Tuberculous meningitis*

Triple therapy with isoniazid, rifampicin and ethambutol; pyrazinamide may be added to this regimen. Treatment should be continued for a year.

5. *Neonatal meningitis*

Most often due to coliform bacilli: give either: (i) gentamicin or (ii) chloramphenicol. Gentamicin penetrates the blood-brain barrier poorly and may be given intrathecally or intraventricularly as well as systemically: often administered in combination with ampicillin.

Group B β-haemolytic streptococcal meningitis in babies is best treated with penicillin.

6. *Meningitis of unknown cause*

When many polymorphs are present but no bacteria have been detected in the deposit of the CSF and there is no growth on culture, treatment must be empirical. This state of affairs is usually the result of inadequate treatment given outside hospital before the patient is admitted. Broth cultures may give a positive result after a few days' incubation when cultures on solid media remain negative.

On a *best-guess basis* give either: (i) a single drug — ampicillin or chloramphenicol — or (ii) triple therapy of penicillin, sulphonamide and chloramphenicol.

35

Sepsis

The term *sepsis* covers a group of numerous and diverse diseases, some trivial and others serious: they include superficial skin infections, wound infection, peritonitis and abscesses. These diseases are amongst the most common encountered in medicine and dealt with in bacteriology laboratories. The causes are also numerous and involve a large number of different bacteria.

SKIN INFECTION

The skin is an efficient barrier to infection and, provided that it is not breached, usually prevents invasion either by resident commensal or exogenous bacteria. Nevertheless skin infections are common probably because minor skin trauma is a part of everyday life.

The main forms of skin sepsis are shown in Table 35.1.

Table 35.1   Skin infections

Infection	Site	Causal organism
Boil	Hair follicle	*Staph. aureus*
Carbuncle	Multiple hair follicles	*Staph. aureus*
Stye	Eyelash follicle	*Staph. aureus*
Sycosis barbae	Shaving area	*Staph. aureus*
Impetigo	Cheeks, around mouth	*Staph. aureus* *Strep. pyogenes*
Erysipelas	Face, sometimes limbs	*Strep. pyogenes*
Pemphigus neonatorum	Infant's skin	*Staph. aureus*
Toxic epidermal necrolysis	Infant's skin	*Staph. aureus*
Acne vulgaris	Face and back	*P. acnes*

CLINICAL FEATURES

Boils, carbuncles, styes, sycosis barbae

With the exception of carbuncles, these skin infections are uncomfortable and unsightly rather than serious. Carbuncles are rarely seen nowadays except in diabetics who have a predisposition to develop septic lesions; they are associated with considerable malaise and constitutional disturbance.

Below are listed some characteristics of these forms of sepsis.

1. Due to *Staphylococcus aureus* mainly of phage groups I or II.
2. Tend to be recurrent — appearing in crops often over weeks or months.
3. Infection is usually endogenous due to a strain carried in the nose and the skin.
4. Generally commoner in males than females: surprisingly, often seen in healthy young males; sycosis barbae is a chronic infection seen only in males (Fig. 35.1).

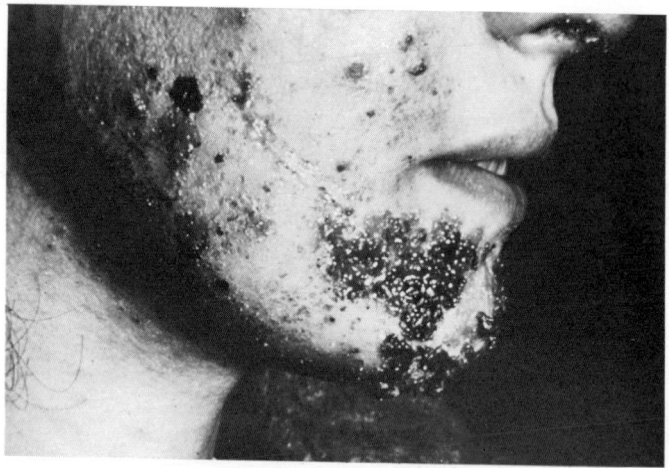

Fig. 35.1 Sycosis barbae. Staphylococcal infection of the skin of the shaving area. (Photograph by Dr A. Lyell.)

Treatment: boils and styes do not require antibiotic therapy and, in any event, this does not prevent recurrences. Severe infections like carbuncles should be treated with penicillin if the causal strain is sensitive; otherwise with a different antistaphylococcal drug such as flucloxacillin. Sycosis barbae should be treated with topical antibiotics e.g. a cream containing neomycin and bacitracin or fusidic acid.

Carriage sites: application of creams containing antiseptics, e.g. chlorhexidine or antibiotics such as neomycin or bacitracin to the nostrils suppresses (but rarely eradicates) nasal carriage of *Staph. aureus;* regular use of hexachlorophane soap reduces overall skin carriage.

Impetigo
A disease of young children in which vesicles appear on the skin around the mouth later to become purulent with characteristic honey-coloured crusts: nowadays most often due to *Staph. aureus* but *Streptococcus pyogenes* can also cause impetigo.

Outbreaks are not uncommon in schools with infection spread by contact or via shared towels and other contaminated fomites.

Glomerulonephritis has been described following impetigo due to *Strep. pyogenes* especially Griffith type 49.

Treatment: antibiotics topically, e.g. tetracycline, chloramphenicol, bacitracin; systemic antibiotics are rarely required for this superficial infection. *Note.* Dermatologists avoid the local application of penicillins which are liable to cause hypersensitivity.

Erysipelas
A spreading infection due to *Strep. pyogenes* presenting as a red, indurated and sharply demarcated area of skin: oedema often develops causing a characteristic 'orange skin' texture to the skin; the patient may be acutely ill with high fever and toxaemia. A rare disease nowadays.

Treatment: penicillin.

Neonatal skin sepsis
Neonates are very susceptible to infection and *Staph. aureus* can become epidemic in neonatal nurseries with outbreaks of pustules, sticky eyes, boils and abscesses. Neonates readily become carriers in nose, skin and umbilical stump.

Two severe infections associated with skin splitting are sometimes seen in neonates and infants; both are due to staphylococci of phage group II which produce an epidermolytic toxin which causes skin splitting and desquamation.

1. *Pemphigus neonatorum:* in which the skin splitting is localised in large vesicles or bullae; although a more serious disease than the neonatal infections listed above, it responds well to antibiotic treatment.
2. *Toxic epidermal necrolysis,* also called *Ritter-Lyell disease* or, more descriptively, 'scalded skin syndrome': a serious disease in which large areas of the skin desquamate leaving a red weeping surface which resembles a scald. Seen predominantly in neonates but also

in young children and occasionally in adults: usually responds to antibiotic treatment with good recovery.

Treatment: flucloxacillin.

DIAGNOSIS OF SKIN INFECTIONS

Specimens: swabs from lesions: pus, exudate.

Direct Gram film: observe for bacteria, noting especially the arrangement of any Gram-positive cocci. (Figs 35.2 and 35.3).

Culture: blood agar, (i) aerobically and (ii) anaerobically — for enhanced growth with better demonstration of β haemolysis by *Strep. pyogenes*.

Observe: for typical colonies.

Identify: as appropriate for the organisms isolated.

Acne vulgaris

A common and disfiguring skin disease of adolescence: sometimes persists into adult life leaving residual pitting or scarring; probably not primarily an infectious disease but bacteria seem to play some role in its pathogenesis.

Propionibacterium acnes (and *P. granulosum*) can regularly be isolated from inflamed comedos (whiteheads): these bacteria probably induce an inflammatory reaction in the skin by the production of lipase which liberates irritant fatty acids from lipid in sebum within sebaceous glands.

Treatment: long-term tetracycline in severe cases.

Note: Bacteriology plays no part in diagnosis.

CELLULITIS

An infection of the subcutaneous tissue: there are two main clinical — and bacteriological — forms:

1. *Acute pyogenic cellulitis* due to *Strep. pyogenes:* presents as a red painful swelling usually of a limb and commonly associated with lymphangitis and lymphadenitis involving local draining lymph glands.

   Treatment: penicillin.

2. *Anaerobic cellulitis:* a rare infection sometimes due only to non-sporing anaerobes or clostridia but more usually a synergistic infection with both aerobic and anaerobic bacteria; the aerobes produce reducing or anaerobic conditions which enable the anaerobes to multiply.

   Causal organisms: a combination of *aerobes* (coliforms, *Pseudomonas*

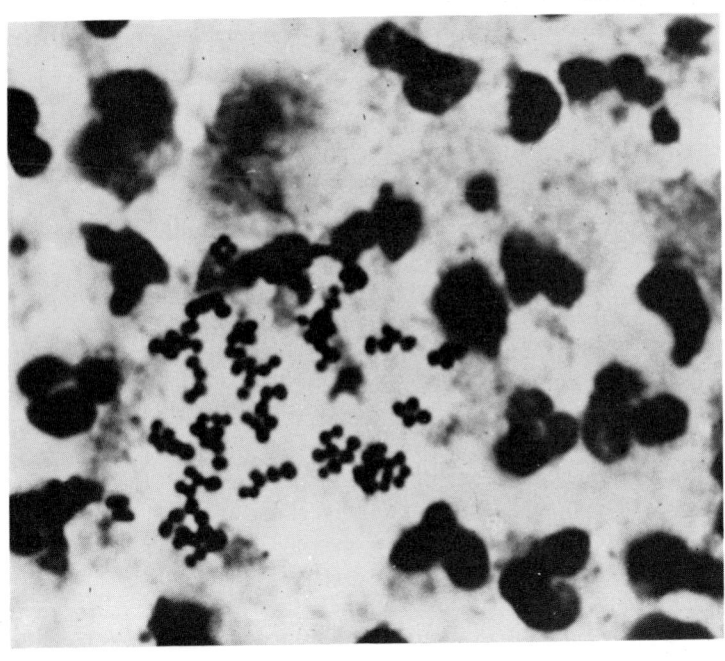

Fig. 35.2 Staphylococcal pus. The cocci are arranged in clusters (magnified ×
4000).

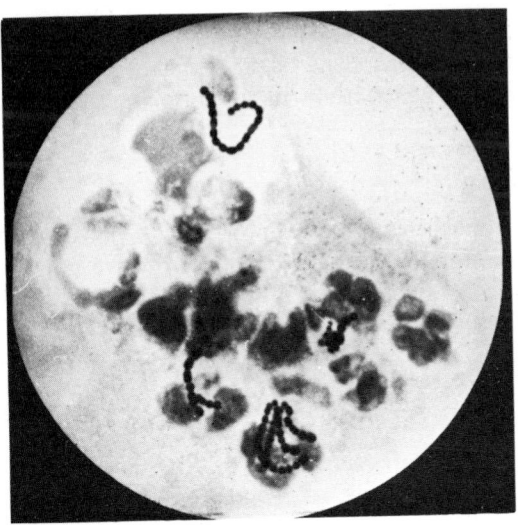

Fig. 35.3 Streptococcal pus. The cocci are arranged in chains (magnified ×
2000).

aeruginosa, Staph. aureus, Strep. pyogenes) and *anaerobes* (most often bacteroides, anaerobic cocci — rarely clostridia).

Clinically: there is redness, swelling and oedema, i.e. a non-specific cellulitis around or near to the primary wound or area of sepsis which is usually situated on the abdomen, buttock or perineum, less commonly on the leg. Two syndromes may develop:

a. *Progressive bacterial gangrene:* in which the skin is rapidly involved, becomes purple in colour and develops central necrosis. When this condition affects the penoscrotal region it was formerly called 'Fournier's gangrene'.

   Treatment: incision and appropriate antibiotics.

b. *Necrotising fasciitis (Meleney's gangrene):* in which the external appearance of the skin remains normal initially while the infective-ischaemic process spreads along the fascial plane causing extensive necrosis. Later the overlying skin, deprived of its blood supply, becomes painful and red and, finally, numb and necrotic. The patient is severely ill with fever and toxaemia, sometimes with septic shock.

   Treatment: wide excision of the skin to expose the entire area of necrotising fasciitis and general supportive measures including appropriate antibiotics.

WOUND INFECTION

Minor wounds are commonplace in everyday life and, naturally, some become infected. *Staph. aureus* is the commonest cause of infections of this sort most of which heal without antibiotic treatment.

HOSPITAL WOUND INFECTION

Clinically, a much more serious problem. Obviously the majority of *surgical patients* have wounds and these involve not only skin and subcutaneous tissue but also muscle and deeper tissues including sometimes bone, peritoneum and viscera.

CLINICAL FEATURES

Surgical wound infection

Presents as a reddening of wound edges with pus formation: sometimes the pus gathers below the suture line in the deeper layers of the wound to form a *wound abscess;* as a rule this eventually discharges to the surface through the sutured incision.

Patients may develop some fever but constitutional disturbance is often minimal: the main problem is that infection impairs and delays healing and therefore prolongs hospitalisation.

Complications:

1. *Dehiscence:* the wound may break down completely — sometimes with exposure of viscera, etc. — and require to be resutured.
2. *Spread of infection to:*
   a. *local tissues* — e.g. the peritoneum in the case of abdominal wounds.
   b. *the blood* — causing septicaemia.

The main bacterial causes of surgical wound infection are shown in Table 35.2.

Table 35.2 Main bacterial causes of surgical wound infection

Bacteria	Species	Most common site
1. Aerobic	*Staph. aureus*	Any wound
	Esch. coli	Any wound but
	Proteus species	especially abdominal,
	Klebsiella species	urological and
	Strep. faecalis	gynaecological
	Ps. aeruginosa	Urological; burns
2. Anaerobic	*B. fragilis*	Abdominal and
	B. melaninogenicus	gynaecological
	Anaerobic cocci	
	Cl. welchii	

Source: most wound infections are due to *endogenous infection*, e.g. with *Staph. aureus* carried in the patient's nose or skin, or bacteria such as coliforms or bacteroides present in the gut as commensals. During hospitalisation before surgery the patient may become colonised with antibiotic-resistant bacteria from the environment and these may later be responsible for wound infection: this represents endogenous infection from a hospital source.

Pseudomonas aeruginosa is an exception to this: rarely found in the human bowel it is most often acquired *exogenously* from contamination of the environment, e.g. water, fluids, ventilators, humidifiers, etc.

Exogenous infection of wounds is also sometimes acquired during operation from the surgeon or other theatre personnel.

Mixed infection: surgical wounds are often infected with more than one bacterial species (except in the case of *Staph. aureus* which is usually present in pure culture): for example, it is common to find more than one species of coliform together with *Strep. faecalis* and

Bacteroides species or *Clostridium welchii* this flora is known to bacteriologists as *faecal flora* and is the usual cause of abdominal wound sepsis. Note that the presence of *Cl. welchii* does not imply that gas gangrene is about to supervene. The pathogenic potential of *Bacteroides species* and anaerobic cocci was underestimated in the past: their establishment in a wound may be aided by the presence of coliforms which use oxygen to promote anaerobic conditions and also by blood and damaged tissue.

Factors affecting surgical wound sepsis
The following factors increase the likelihood of wound infection:
1. Operations involving opening of the bowel.
2. Presence of drains or other foreign bodies.
3. Long operations.
4. Large wounds with considerable tissue trauma.
5. Obesity.

Prevention
Both difficult and complex: rigid observance of aseptic and antiseptic technique both in preparation for and during operation; many patients are colonised before operation with the organisms which subsequently infect their wounds — the risk of this is increased by length of stay in hospital.

Theatre: surgeons and assistants must wear gowns to prevent bacterial dispersal via squames shed from the body surface and masks to trap respiratory secretions; surgeon's hands and patient's skin must be disinfected before operation; the surgeon and his assistants must wear gloves; theatres should be kept free from dust and have positive pressure ventilation to prevent air and dust being sucked in from outside; they should be sited away from main thoroughfares in the hospital.

Infection is most often acquired at operation by bacteria entering the wound either from the patient's commensal flora or from the skin flora of the surgeons and assistants.

Ward: bacteria abound in hospital wards: patients with discharging wounds should be isolated to prevent the dissemination of pathogenic bacteria; floors must be kept clean and free from dust; blankets, now made from cellular cotton and not wool, must be regularly washed at a sufficiently high temperature to kill vegetative bacteria.

Antibiotic prophylaxis: if carefully chosen, antibiotics given peroperatively and for a short time thereafter can reduce the incidence of wound infection after certain operations; especially indicated when heavy soiling of the wound edges is unavoidable, e.g. in colonic surgery.

SPECIAL TYPES OF WOUND INFECTION

1. **Burns**

Extensive burns consist of a large moist exposed surface which is usually being prepared for a skin graft; always heavily colonised with bacteria — surprisingly this often does not prevent the graft taking. However, *Strep. pyogenes,* now a rare but still a highly dangerous organism in a burns unit, not only affects the patient's general well-being, e.g. by causing septicaemia, but also causes the graft to fail.

Pseudomonas aeruginosa has a particular propensity to persist in a burns unit: of relatively low invasive powers, it can be exceedingly difficult to eradicate both from an individual patient and from the environment of the unit.

Septicaemia: the invasion of the blood stream with bacteria from the burn is a dangerous complication.

2. **Orthopaedics**

Wound infection is also a problem in orthopaedic surgery where healing of bone by first intention is important for future weight-bearing and movement. *Staph. aureus* is the principal cause.

Hip replacement has revolutionised the mobility of elderly patients with arthritis: about 4 per cent of these operations fail due to low-grade chronic infection which causes the prosthesis to work loose; most often due to *Staph. epidermidis.*

3. **Puerperal sepsis** (including septic abortion)

Formerly due mainly to *Strep. pyogenes* the important causes nowadays are streptococci of Lancefield group B, bacteroides (especially *B. melaninogenicus*) and anaerobic cocci; still a severe disease which requires prompt treatment.

Cl. welchii is a very rare but dangerous cause of puerperal sepsis.

4. **Tracheostomies**

Relatively common in modern hospitals — especially with the introduction of intensive-care units; the wounds have a marked tendency to become colonised with coliforms — which are often multiply-antibiotic resistant — from the hospital resident flora. Low-grade pathogens like *Acinetobacter* and *Serratia* species can be particularly difficult to eradicate; it is often uncertain if the presence of these and other coliforms in tracheal secretions actually harm the patient.

DIAGNOSIS OF WOUND INFECTION

Specimen: swab of pus, exudate, tissue from wound.

Direct Gram film: observe for organisms present.

Culture: on a variety of media such as: blood agar, CLED agar and MacConkey agar — aerobically; blood agar and aminoglycoside blood agar — anaerobically; Robertson's meat medium.

Observe and identify: the bacteria isolated by standard tests.

TREATMENT

Antibiotics appropriate for the bacteria isolated.

PERITONITIS

A serious complication not only of abdominal surgery but of diseases that may cause spontaneous perforation of the gastrointestinal tract, e.g. peptic ulcer, diverticulitis, acute appendicitis, Crohn's disease: perforated typhoid ulcer is still a cause of death in enteric fever.

CLINICAL FEATURES

The patient's condition deteriorates markedly with fever, toxaemia and shock: there may be tenderness on palpation of the abdomen and absence of bowel sounds due to paralytic ileus.

Causal organisms: usually a mixed infection with 'faecal flora', i.e. coliforms, *Strep. faecalis, B. fragilis.*

DIAGNOSIS

A specimen of peritoneal exudate is examined in the same way as pus from a wound.

TREATMENT

Appropriate antibiotics: it is usually necessary to start therapy before sensitivity results are available; a suitable combination would be:

  Ampicillin
  Gentamicin
  Metronidazole
  Alternative drugs: cefuroxime or cefoxitin

Therapy should be reviewed as soon as the results of sensitivity tests are known.

CLOSTRIDIAL WOUND INFECTION

Wound infection primarily due to clostridia differs clinically from the infections described above: clostridial anaerobic wound infection is both rarer and much more severe.

The two main diseases associated with clostridial wound infection are shown in Table 35.3.

Table 35.3   Clostridial wound infection

Disease	Causal organisms
1. Gas gangrene	*Cl. welchii* (65% of cases) *Cl. oedematiens* (20–40% of cases) *Cl. septicum* (10–20% of cases)
2. Tetanus	*Cl. tetani*

Neither disease can really be considered as 'sepsis' since they do not form pus but can conveniently be discussed under the heading of wound infection.

Gas gangrene
A rare disease in peacetime but a scourge amongst the wounded of armies in the field — notably during the First World War.

CLINICAL FEATURES

A spreading gangrene of the muscles with profound toxaemia and shock: there is oedema with blackening of the tissues and a foul-smelling serous exudate; crepitus can often be detected under the skin due to gas production by the clostridia.

Severity: a serious disease with a high case fatality rate.

PATHOGENESIS

Due to contamination of wounds by dirt and soil containing clostridia derived from animal faeces: infection is favoured by severe, extensive wounding with the presence of necrotic tissue, blood clot and foreign bodies — all of which produce anaerobiasis. Distant vascular damage may impair blood supply to the site.

Wounds: apart from war injuries, gas gangrene is encountered in civilian practice following major trauma; very occasionally it complicates operations on the bowel (*Cl. welchii* is a normal bowel inhabitant), mid-thigh amputation or vascular surgery on the ischaemic limb.

Clostridia: more than one species is often present in gas gangrene.

Toxins: clostridia produce powerful toxins which themselves cause tissue danage — and so anaerobiasis — and thus enhance spread of the infection.

DIAGNOSIS

Specimen: exudate, tissue.

Direct Gram film: observe for typical Gram-positive bacilli — spores may or may not be seen. In clinical material *Cl. welchii* is usually capsulated but it does not form spores.

Culture: blood agar and aminoglycoside blood agar anaerobically: Robertson's meat medium.

Observe: for typical colonies.

Identification: Cl. welchii — Nagler reaction; identification of other clostridia may be difficult — it is largely based on biochemical reactions and toxin production.

TREATMENT

Surgical: by wide excision or amputation of affected tissue.

Antibiotics: large doses of penicillin; perhaps with metronidazole in addition.

Antitoxin: widely used in wartime; of doubtful value.

Hyperbaric oxygen to reduce anaerobiasis in tissues has been reported to be effective. This supportive treatment requires special apparatus: if available it should be administered.

Tetanus

A classic example of an exotoxic bacterial disease: due to *Cl. tetani:* the toxin produced is one of the most powerful known.

Tetanus is a very rare disease in the UK but the wounds from which it may develop are common and may be trivial.

CLINICAL FEATURES

Severe and painful muscle spasms: the masseter muscles are often affected early in the disease causing 'lockjaw' — the familiar name for tetanus — and 'risus sardonicus' — the characteristic facial grimace.

Because the extensor muscles of the body are more powerful than the flexor muscles, as the spasms progress the body becomes arched in 'opisthotonus' with only the patient's head and heels touching the bed. Death is due to exhaustion, asphyxiation or intercurrent infection.

PATHOGENESIS

Cl. tetani produces a protein exotoxin — mainly released by bacterial lysis. It has two components:

1. *Tetanospasmin,* which acts on synapses to block the normal inhibitory mechanism which controls motor nerve impulses.
2. *Tetanolysin:* lyses erythrocytes.

Spread: Cl. tetani does not spread beyond the wound but the toxin, absorbed at the motor nerve endings, travels via the nerves to the anterior horn cells in the spinal cord.

Site: wounds of the face, neck and upper extremities are more dangerous than those of the legs and feet: they are associated with more severe disease and a shorter incubation period (usual range for tetanus is 5–15 days).

EPIDEMIOLOGY

Source: animal faeces, *Cl. tetani* is hardly ever found in human faeces. Wounds become contaminated with spores and, if anaerobic conditions are present, the spores germinate to produce vegetative bacilli which form toxin.

Worldwide in distribution but the incidence is very much higher in the Third World — where it is an important cause of death — than in the developed countries.

Tetanus neonatorum, in which the umbilical stump is the portal of entry, is still common in rural areas of Asia, Africa and South America.

Classically tetanus is associated with severe wounds contaminated with soil or dust similar to those that precede gas gangrene. Today in countries with good medical services, prophylactic measures prevent patients with such wounds developing the disease and tetanus often follows minor injuries disregarded by the patient, e.g. a small penetrating wound from a splinter of wood. In fact, in some of the cases of tetanus in the UK no wound can be found.

DIAGNOSIS

The diagnosis is clinical: attempts at bacteriologal confirmation often fail.

Specimen: swab or exudate from wound.

Direct Gram film: examine for characteristic Gram-positive bacilli with round terminal spores ('drum-sticks').

Culture: blood agar and aminoglycoside blood agar anaerobically; Robertson's meat medium.

Observe: typical translucent spreading colonies; subculture from the edge to obtain a pure growth.

Identify: by Gram film, biochemical tests and demonstration of the exotoxin (provisionally by inhibition of haemolysis on blood agar by specific antitoxin; confirmation by demonstration of mouse pathogenicity and its prevention by antitoxin).

TREATMENT

Supportive: artficial ventilation with muscle relaxants to control spasms.

Antitoxin: large doses intravenously to neutralise toxin.

Excision of wound.

Antibiotics (penicillin or tetracycline): to prevent further toxin production.

PREVENTION

Official policy in Britain is for active immunisation in childhood with formol toxoid as part of the Triple Vaccine.

In casualty departments: a common problem is the wound contaminated with soil or dust and therefore, potentially, with *Cl. tetani.* The wound must first be thoroughly cleansed and then:

(i) If the patient has been previously fully immunised a booster dose of tetanus toxoid should be given; if the wound is dirty and more than 24 h old human antitetanus immunoglobulin must be administered in addition.

(ii) If the patient is non-immune or if the previous history of active immunisation is uncertain give human antitetanus immunoglobulin as well as starting a full course of tetanus toxoid by injection at another site.

Patients or the parents of an injured child are often unable to give a history relating to tetanus vaccination. Parents may know that their child has been immunised against the infectious diseases diphtheria and pertussis without realising that the triple vaccine also protects against tetanus.

Penicillin (usually a mixture of short and long acting varieties) has been recommended on its own as prophylaxis but its value is not proven. However, in casualty practice it is common to give penicillin to the wounded not only as part of the scheme to prevent tetanus but also in an attempt to avoid pyogenic infection.

ABSCESSES

A most important form of sepsis — a threat to health and recovery and sometimes very difficult to diagnose. Abscesses, both obvious and cryptogenic, are an important part of the work of hospital bacteriology laboratories.

An abscess is a collection of pus within and often deeply within the body: abscesses are walled off by a barrier of inflammatory reaction with fibrosis. It is therefore often impossible to treat them satisfactorily by antibiotics alone — surgery and drainage are also necessary.

Abscesses can form in almost any tissue or organ of the body but some sites are much more common than others: these are listed in Table 35.4.

Table 35.4 Abscesses — some common sites

Site	Route of Infection — predisposing factors	Bacteria usually responsible
Subcutaneous tissues e.g. finger pulp, palmar space	Penetrating wounds	*Staph. aureus,* occasionally *Strep. pyogenes*
Axilla	Extension of superficial infection or via lymphatics that involve lymph nodes with suppuration	
Sub-mandibular		
Breast	Breast-feeding — infected from infant	*Staph. aureus*
Peritonsillar (quinsy)	Streptococcal sore throat	*Strep. pyogenes*
Intraabdominal e.g. appendix, subphrenic, paracolic	Appendicitis Abdominal sepsis-peritonitis due to any cause	Faecal flora, *Strep. milleri*
Ischiorectal	Direct from rectum	Faecal flora
Perianal	Infected hair follicle round anus	*Staph. aureus,* faecal flora
Pelvic	Abdominal, gynaecological sepsis	Faecal flora, Genital flora
Tubo-ovarian (pyosalpinx)	Gynaecological sepsis Gonorrhoea	Genital flora, *N. gonorrhoeae*
Bartholin's gland	Local spread	Genital flora, *N. gonorrhoeae*
Perinephric	Extension of acute pyelonephritis	Coliforms
Cerebral	Otitis media; sinusitis Haematogenous	*Strep. milleri,* Bacteroides *Staph. aureus,* *Proteus species*
Hepatic	Ascending cholangitis Portal pyaemia	Faecal flora, *Strep. milleri*
Lung	Aspiration pneumonia Bronchial obstruction with collapse (e.g. tumour) *Staph. aureus* pneumonia	Oro-pharyngeal flora *Staph. aureus*

Note: the anaerobic members of the commensal flora (*Bacteroides species*, anaerobic cocci, etc.) play an important role in abscess production.

Note. Tuberculosis commonly causes 'cold' abscesses but these are discussed in Chapter 38 and are not included here. Cold infers that these abscesses are not accompanied by an acute inflammatory reaction.

CLINICAL FEATURES

Abscesses may be: (i) clinically obvious or (ii) cryptogenic.

Clinically obvious abscesses include those at sites such as breast, axilla, peritonsillar (quinsy), perianal, ischiorectal and Bartholin's glands: there is painful swelling with local inflammation, fever and often some degree of constitutional upset: brain abscess commonly presents with the signs of a space-occupying lesion most often in the temporal lobe (Fig. 35.4).

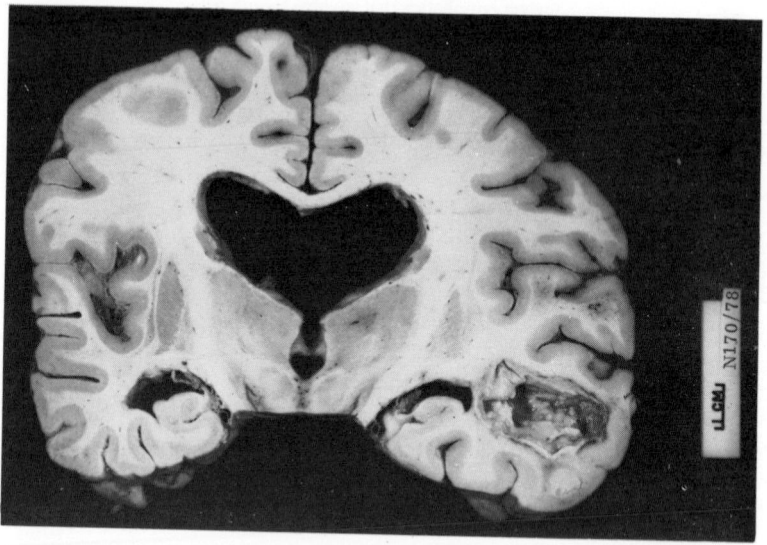

Fig. 35.4 Cerebral abscess. Section of brain with an abscess in the right temporal lobe secondary to suppurative otitis media. (Photograph by Professor J. Hume Adams.)

Cryptogenic abscesses can be exceedingly difficult to diagnose and, unfortunately, are not uncommon: many abdominal and pelvic abscesses are of this type — the patient presents with vague, progressive ill health and fever: although obviously toxic there are no definite localising signs.

In addition to the clinical history there may, however, be other clues: for example, a subphrenic abscess may be detected radiologically: a positive blood culture substantiates sepsis although it does not localise it: in some cases laparotomy will be required if there is a high degree of clinical suspicion.

PATHOGENESIS

Abscesses are localised infections but do not necessarily form at the site of primary infection: there may be tracking of pus with its collection at a site some distance from the original infection; subphrenic and pelvic abscesses are examples of this.

Metastatic abscesses: occasionally, abscesses become multiple usually as a result of blood-borne or 'pyaemic' spread of infected thrombi; such abscesses may form in many sites. *Portal pyaemia* in which the source is intraabdominal sepsis (often the appendix) is an example of this which results in liver abscesses.

DIAGNOSIS

Specimen: pus (which may have to be collected at operation or by aspiration); blood culture.

Culture: on blood agar aerobically and anaerobically, Robertson's meat medium; incubation may have to be continued for some days.

Observe: for bacterial colonies or growth.

Identify: by tests appropriate for the bacteria isolated.

TREATMENT

Antibiotics are rarely sufficient on their own: if the abscess does not discharge spontaneously (and this can have serious effects) surgical intervention and drainage may be necessary; this should be done under appropriate antibiotic cover; the drugs to be administered depend on the site of the abscess and, therefore, the likely infecting bacteria (Table 35.4).

36

Pyrexia of unknown origin

Definition. Popularly known as PUO, this clinical entity has been defined in a number of different ways; basically, the patient must have a fever that is

(i) significant (a temperature over 37.5°C)
(ii) persistent (for at least 1 week: usually longer, often 3 weeks)
(iii) without a readily identifiable cause (i.e. no localising signs or symptoms)

Cause: is more often a relatively common disorder with an atypical presentation than an obscure disease.

Diagnosis: can be exceedingly difficult; in some cases it is never made, and the patient may recover spontaneously.

The principal types of disease which can be responsible for PUO are shown in Table 36.1

Table 36.1 Causes of pyrexia of unknown origin

Cause		Percentage of cases
Infections		40
Neoplasms	Especially lymphoma, leukaemia, but also other forms of cancer, e.g. hypernephroma, disseminated malignancy	20
Connective tissue diseases	Systemic lupus erythematosus, polyarteritis nodosa, temporal arteritis	20
Others, e.g.	(i) Granulomatous diseases — Crohn's disease, sarcoidosis (ii) Drug-induced fevers (iii) Malingering ("factitious fever")	20

INFECTION

The most important cause of PUO: with a stringent definition, infection accounts for 40 per cent of cases; with less rigid criteria (i.e.

234

lower grade fever of shorter duration) it is responsible for at least 70 per cent of cases.

Bacterial infections which can present as PUO are listed in Table 36.2.

Table 36.2 Bacterial diseases commonly presenting as pyrexia of unknown origin

Cause	Disease — special features
Systemic infections	
Infective endocarditis	Infection of abnormal heart valves
Tuberculosis	Pulmonary, non-pulmonary (e.g. bone, renal) or cryptic miliary
Enteric fever	Usually acquired abroad
Brucellosis	Occupational association
Leptospirosis	Occupational association
Localised sepsis	
Hepato-biliary sepsis	Cholecystitis, cholangitis, liver abscess
Intra-abdominal abscess ⎫ Pelvic abscess ⎬	Usually follow intestinal or gynaecological sepsis; sometimes post-operative
Renal infections	Chronic pyelonephritis; perinephric abscess
Sinusitis	
Dental infection	Apical dental abscess

Other infectious causes of PUO include:

Viral diseases endemic in the UK: infectious mononucleosis (glandular fever), hepatitis A: others such as enterovirus, cytomegalovirus infections, childhood fevers with atypical presentation.

Q fever: due to a rickettsia-like organism, *Coxiella burneti;* most common in males in agriculture.

Psittacosis: due to *Chlamydia psittaci;* suspect if the patient keeps parrots, budgerigars or pigeons.

Protozoal diseases: malaria, amoebiasis (with liver abscess), visceral leishmaniasis (kala-azar), trypanosomiasis — all of which are tropical infections; toxoplasmosis, worldwide in distribution, is usually acquired in this country.

Imported diseases: the common causes of PUO in the UK are very different from those seen abroad. Air travel has allowed the importation of tropical diseases within their incubation periods so that they may present in this country as undiagnosed fevers. It is essential to take an accurate 'geographical history' from all patients. Malaria is the most important disease in this category; almost all cases of typhoid fever in Britain have been acquired abroad.

Lassa fever, Ebola fever and Marburg disease are viral diseases imported from Africa — all rare and very serious; once suspected, patients require to be investigated and nursed in strict isolation because of the danger of infecting nursing and other staff in attendance.

INVESTIGATION

This can be both time-consuming and difficult: below are the main examinations carried out; emphasis has been placed on those that help to establish an infective cause for PUO.

1. *Clinical history:* to include patient's previous illnesses, family illnesses, foreign travel, contact with pets or farm animals, occupation, currently prescribed drugs

2. *Physical examination:* especially to detect lymphadenopathy or enlargement of liver or spleen.

3. *Laboratory examinations:*
   a. *Bacteriology*
   Blood culture: serial cultures should be taken: isolation of viridans streptococci points to infective endocarditis: isolation of coliforms, *Streptococcus faecalis, Strep. milleri,* or *Bacteroides species* suggests the possibility of intra-abdominal sepsis.
   Urine microscopy and culture: serial specimens may be necessary to detect intermittent bacteriuria in chronic pyelonephritis; microscopic haematuria is often present in infective endocarditis and hypernephroma.
   Stool culture and microscopy: indicated if there is a history of diarrhoea; again several specimens should be examined.
   Serology: appropriate tests for enteric fever (Widal reaction), brucellosis, infectious mononucleosis, Q fever, leptospirosis, and for the other infections listed above.
   b. *Haematology*
   Full blood count: blood films with differential white cell count and examination for malarial parasites: erythrocyte sedimentation rate — a very high result (e.g. more than 100 mm per h.) suggests a connective tissue disease or malignancy.
   c. *Biochemistry*
   Liver function tests, etc.

4. *Radiological examinations:* straight X-rays of chest, abdomen, sinuses and teeth.

Further investigations should be planned at this stage if the diagnosis has not been made. The order in which they are carried out depends on what is considered the most likely cause of the PUO. Examples are listed below:

1. *Serological tests* for connective tissue diseases; e.g. antinuclear factor, anti-native DNA and other autoantibody tests. Immunoglobulin studies in myeloma.
2. *Specialised radiology and ultrasonography:* to identify the site of disease, e.g. cholecystogram, excretion urogram, barium studies (in inflammatory bowel disease), isotope and, if available, CAT scanning, lymphangiogram (in lymphoma).
3. *Biopsy:* of lymph nodes, bone marrow, muscle, liver, kidney.
4. *Laparotomy:* a last resort if there is evidence of intra-abdominal pathology.

37

Septicaemia and endocarditis

Two different but sometimes interrelated diseases: both are serious infections with a considerable mortality.

SEPTICAEMIA

Literally 'sepsis of the blood': in practice two terms are used to describe the presence of organisms in the blood stream.

1. *Septicaemia:* when the patient shows evidence of sepsis; a severe disease with considerable morbidity and mortality.
2. *Bacteraemia:* term used if the patient shows no signs of clinical sepsis; often asymptomatic and transient, and usually of minor or no clinical relevance.

Septicaemia is of two kinds:

a. A basic feature of some generalised or 'septicaemic' infectious diseases, e.g. brucellosis, enteric fever.
b. A complication of more localised infections, e.g. pyelonephritis, peritonitis, cholangitis, meningitis, pneumonia, osteomyelitis, abscesses of internal organs, etc.; organisms spread from the focus of infection into the blood stream.

Nowadays the second form is much commoner and is the type that will be described in this chapter: septicaemia as part of generalised infections is described with the diseases concerned.

CAUSAL ORGANISMS

The organisms that cause septicaemia and the infections from which they originate are shown in Table 37.1.

Note: that infective endocarditis is also associated with septicaemia but is described separately and in more detail below.

238

Table 37.1   Common causes of septicaemia

Predisposing factor	Causal organisms
Abdominal sepsis (peritonitis, hepatobiliary infection, abscess, etc)	Coliforms *B. fragilis* *Strep. faecalis* *Strep. milleri*
Infected wounds, burns, pressure sores	*Staph. aureus* *Strep. pyogenes* Coliforms *B. fragilis*
Gynaecological sepsis (puerperal infection, pelvic abscess, salpingitis etc.)	Coliforms *Strep. faecalis* *Bacteroides species* *Strep. pyogenes* Lancefield group B streptococci
Urinary tract infection	Coliforms *Strep. faecalis*
Osteomyelitis Septic arthritis	*Staph. aureus*
Pneumonia	*Strep. pneumoniae*
Meningitis	*Strep. pneumoniae* *N. meningitidis* *H. influenzae*
Meningitis in neonates	Coliforms Lancefield group B streptococci
Drip sites, shunts, intravascular catheters	*Staph. aureus* *Staph. epidermidis* Coliforms
Splenectomised patients	*Strep. pneumoniae*
Immunosuppressed patients	Coliforms *Staph. aureus* *Ps. aeruginosa* *Strep. pneumoniae*

CLINICAL FEATURES

Signs and symptoms: variable — sometimes minimal, sometimes severe and rapidly progressive; the predominant signs may be those of the underlying disease (e.g. pneumonia, peritonitis, etc.). The presenting feature is usually a worsening of the patient's condition with fever, rigors, tachycardia, tachypnoea, cyanosis, hypotension; in elderly patients there may be confusion, agitation or behavioural changes.

Septic shock

Septic shock is a special form of septicaemia in which there is a sudden and catastrophic deterioration in the patient's condition: it is also known as *endotoxic or bacteraemic shock*.

Clinically: shock is characterised by a progressive cardiovascular collapse, often sudden in onset, with severe hypotension and tissue anoxia leading to multiple organ failure (e.g. heart, lungs, liver, kidneys).

Causes: complex; septic shock is usually a complication of septicaemia with Gram-negative bacilli (e.g. after abdominal or urinary tract infection) — and very occasionally of Gram-positive bacterial infection; the mechanism is probably due to circulating bacterial endotoxin which activates the complement system with the release of vasoactive substances and causes intravascular coagulation.

Endotoxin assay: endotoxin can be assayed in the blood by testing for coagulation of an extract prepared from the amoebocytes (blood cells) of the Limulus crab; this test is both specific and extremely sensitive.

DIAGNOSIS

Blood culture

More than one culture may be required: when septicaemia is suspected in a severely ill patient, two separate cultures should be taken from different veins over a period of about 5 minutes: whenever possible before antibiotics are administered.

Observe: for early signs of growth.

Subculture: to blood agar aerobically and anaerobically.

Identify: as appropriate for the organism isolated.

TREATMENT

Control of infection may need surgical intervention (e.g. to drain an abscess, resuture a ruptured viscus, etc.). Antimicrobial therapy is also required and this should be:

1. Bactericidal.
2. Administered intravenously.
3. Prompt and in adequate dosage.

Septic shock needs special resuscitative measures in addition to antibiotic therapy.

Table 37.2 lists some of the antimicrobial drugs of choice.

Table 37.2 Antimicrobial therapy in septicaemia

Organism	Antimicrobial drugs
Strep. pneumoniae *Strep. pyogenes* *Strep. milleri* *N. meningitidis*	Penicillin
Staph. aureus *Staph. epidermidis*	Flucloxacillin, Fusidic acid
Coliforms	Gentamicin, Cefuroxime
Ps. aeruginosa	Gentamicin, Azlocillin
Strep. faecalis	Ampicillin
Bacteroides species	Metronidazole

INFECTIVE ENDOCARDITIS

An infection of the endocardium of the heart valves and sometimes of the endocardium around congenital defects.

Infective endocarditis was formerly known as bacterial endocarditis: the change in nomenclature is in recognition of the fact that many types of organism other than bacteria can cause it.

Mortality: before antibiotics were available the disease was always fatal; even nowadays the case fatality rate is around 30 per cent.

CAUSAL ORGANISMS

These are listed in Table 37.3.

CLINICAL FEATURES

There are two clinical forms of the disease:
1. Acute
A rapidly progressive disease due to pyogenic bacteria such as *Staphylococcus aureus, Streptococcus pneumoniae* or *Strep. pyogenes.*

2. Subacute
The commoner and more chronic disease: also progressive but slowly: due to less pathogenic organisms such as viridans streptococci, *Strep. faecalis, Staph. epidermidis, Coxiella burneti,* etc.

Signs and symptoms: classically — fever, malaise, weight loss, cardiac murmur, anaemia, splinter haemorrhages (i.e. under the finger nails), haematuria, petechiae, splenomegaly.

Table 37.3   Causes of infective endocarditis

Organism	Percentage of cases
Bacteria	
Strep. sanguis ⎤	
Strep. bovis	viridans or	
Strep. mutans ⎬ non-haemolytic	60-80
Strep. mitior	streptococci	
Strep. faecalis ⎦	
Staph. aureus	20-30
Staph. epidermidis	
Other bacteria	10
(e.g. corynebacteria,	
Haemophilus species, coliforms)	
Fungi	
Candida albicans	rare
Aspergillus species	
Rickettsia, chlamydia	
C. burneti	rare
C. psittaci	

Nowadays this full-blown syndrome is rare: most patients present with an insidious illness and few or none of the signs listed above: *the laboratory is crucial in making the diagnosis.*

Clinical course: unless adequately treated (and even in some patients apparently so treated) there is progressive damage to the heart valves leading to cardiac failure and death.

PATHOGENESIS

Most patients have pre-existing cardiac disease (but *note:* some do not and in others this is not diagnosed until after infective endocarditis has developed).

Cardiac and other abnormalities which *predispose to infective endocarditis* are listed below:

1. *Rheumatic valvular disease:* e.g. stenosis or incompetence of the mitral and aortic valves following rheumatic fever.
2. *Congenital defects:* e.g. septal defects, patent ductus arteriosus, coarctation of the aorta.
3. *Intracardiac prostheses:* replacement of diseased heart valves with prosthetic valves.
4. *Degenerative cardiac disease in the elderly.*
5. *Drug abuse:* addicts who take drugs intravenously have a high risk of endocarditis — often with atypical clinical features

PATHOLOGY

Thrombi of fibrin and platelets form on damaged endocardium of the mitral, aortic or tricuspid valves. Usually only a single valve is affected: circulating organisms colonise the thrombi and convert them into infected *vegetations;* the end result is destruction of the valve (Fig. 37.1).

Infection of endocardial thrombi results from:

1. *Transient, asymptomatic bacteraemia:* usually with organisms derived from the normal flora and particularly that of the mouth — which explains the frequency of viridans streptococci as a cause of

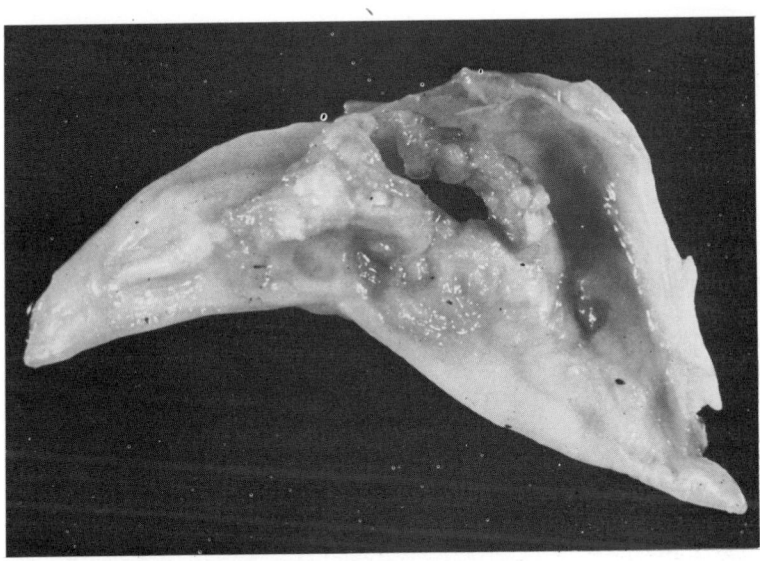

Fig. 37.1   Infective endocarditis. Heart valve from a case of the disease. The valve has a large hole due to the destructive effect of the infection on the tissue of the valve.

    infective endocarditis; this type of bacteraemia is common after dental procedures.
2. *Septicaemia:* rare; usually part of a generalised infection with more virulent organisms (such as *Staph. aureus, Strep. pneumoniae*).

Vegetations: shed organisms into the blood stream — often over long periods of time; minute thrombi are also dislodged to give rise to the distant haemorrhagic and other manifestations of the disease.

Immune complexes: can also form and may produce vasculitis and glomerulonephritis.

Prosthetic valve endocarditis

A rare but usually fatal complication of cardiac surgery: the infection may be acquired at operation when it is usually due to *Staph. aureus, Staph. epidermidis,* yeasts, coliform bacilli, corynebacteria etc.; late-onset endocarditis is acquired in the same way as that of natural valves and the infecting organisms are therefore similar.

DIAGNOSIS

Blood culture

The corner-stone of diagnosis and important for the subsequent treatment of the patient; repeated cultures may be necessary to isolate the causal organism, e.g. two to six, if possible, taken over 48 h.

Observe: for early signs of growth.
Subculture: to blood agar aerobically and anaerobically.
Identify: as appropriate for the organism isolated.

Antibiotic sensitivity tests

The usual disc diffusion method of testing is not adequate in infective endocarditis. The following tests must be carried out:

Minimal inhibitory concentration of potentially useful drugs both singly and in combination against the organism.

Minimal bactericidal concentration must also be determined since it is important to achieve a bactericidal level of antibiotic in the blood (rather than merely a bacteriostatic level).

Culture negative endocarditis

In about 20 per cent of cases no organisms can be grown from blood cultures. This may be due to:

1. Infection with *Coxiella burneti* or *Chlamydia psittaci*; carry out serological tests to confirm or exclude these diagnoses.
2. Recent antibiotic therapy: repeat blood cultures over a few days in absence of chemotherapy.
3. Infection with fastidious organisms difficult to grow in ordinary media (e.g. bacterial L-forms, *Bacteroides species*): repeat blood cultures using special media if these are suspected.
4. The disease is in a phase in which organisms are not being shed into the blood stream: repeat blood cultures.

TREATMENT

Depends on adequate — and this usually means high — dosage with an antimicrobial drug; two drugs in combination are often preferred.

The antibiotic regimen selected must be:

1. Bactericidal.
2. Parenteral (at least initially).
3. Continued for several weeks.

High doses of bactericidal drugs are necessary because the aim of therapy is to eliminate the organisms from their sites enmeshed in the relatively avascular vegetations.

Some of the antibiotic regimens used are shown in Table 37.4.

Table 37.4 Antibiotic regimens for infective endocarditis

Causal organism	Antimicrobial drugs
Virdans streptococci	Penicillin and gentamicin
Strep. faecalis	Penicillin and gentamicin*
Staph. aureus *Staph. epidermidis*	Flucloxacillin and gentamicin, Flucloxacillin and fusidic acid
Coliforms	Ampicillin and gentamicin, Cephalosporin and gentamicin
Fungi	Amphotericin B and 5-fluorocytosine
C. burneti *C. psittaci*	Tetracycline

*Although *Strep. faecalis* is not sensitive to either drug alone, this combination is usually bactericidal and, in practice, effective.
Note: Vancomycin is a useful alternative drug for infections due to streptococci and staphylococci.

Laboratory monitoring of therapy. (i) Estimation of the antibiotic level in patient's serum; rarely necessary if large doses are being given intravenously but may be indicated when there is a switch to oral therapy; also used to avoid overdosage with toxic drugs (e.g. aminoglycosides). (ii) Estimation of the bactericidal activity of the patient's serum against the causal organism: although of little prognostic value, a poor result (e.g. no killing at serum dilution of 1 in 8) indicates that the antibiotics chosen, their dose and route of administration should be reconsidered.

Surgery: replacement of damaged valves is now accepted as part of the management of cases of infective endocarditis; it is often life-saving particularly in cases of Q fever endocarditis.

PROPHYLAXIS

Although not of proven value, 'at risk' patients (e.g. those with valvular disease) should be given prophylactic antibiotics — oral amoxycillin or parenteral penicillin with gentamicin — before dental or other operative procedures.

38

Tuberculosis and leprosy

TUBERCULOSIS

A chronic debilitating disease once the scourge of Victorian Britain ('consumption') which remains a major health problem in much of the Third World today.

Causal organisms: Mycobacterium tuberculosis — the human tubercle bacillus but also *Mycobacterium bovis* — the bovine tubercle bacillus.

CLINICAL FEATURES

Tuberculosis is a slowly progressive, chronic, glanulomatous infection which most often affects the lungs; other organs and tissues may also be involved.

Respiratory (pulmonary) tuberculosis: recognised as: (i) primary infection; (ii) post-primary infection.

Primary infection

Primary tuberculosis of the lung takes the form of the *primary complex* — a local lesion (Ghon focus) with marked enlargement of the regional hilar lymph nodes. The Ghon focus develops at the lung periphery, usually just below the pleura in the midzone, and is a small lesion.

Many cases of primary tuberculosis are entirely symptomless and the patient, a child or young adult, is unaware of the infection: when symptoms are present, they are often vague and non-specific such as malaise, anorexia, weight loss, fever, sweats, tachycardia; cough may not be prominent.

Progressive primary infection

Disseminated tuberculosis: progressive disease may follow primary infection although it is usually contained by the development of hypersensitivity which causes the primary focus to become walled-off and calcified: antibiotic treatment of symptomatic diagnosed primary infection prevents progression. Some forms of tuberculosis that result from spread of primary infection are listed below:

1. *Tuberculous bronchopneumonia:* an acute diffuse involvement of the

lung due to discharge into the bronchial tree of caseous material from an expanding Ghon focus or, more often, from caseous hilar lymph nodes (Fig. 38.1 a and b). A serious and often fatal complication if untreated.

2. *Miliary tuberculosis:* with small tuberculous foci (tubercles) disseminated widely throughout the body as a result of haematogenous (blood-borne) spread of infection.

3. *Tuberculous meningitis:* in which the infection penetrates the blood-brain barrier to involve the meninges; uniformally fatal before the antibiotic era, this remains a serious disease with considerable mortality and disabling sequelae in some survivors.

4. *Bone and joint tuberculosis:* can affect many different sites: a common form is spinal tuberculosis in which there may be collapse of the vertebrae causing kyphosis and formation of a psoas abscess in the groin due to tracking of pus down the psoas muscle from the infective process in the spine.

5. *Genitourinary tuberculosis:*
   a. *Renal tuberculosis* presents with frequency and painless haematuria: the urine shows a 'sterile' pyuria — numerous pus cells on microscopy but no growth of pathogens when cultured on standard media.
   b. *Endometrial tuberculosis* in females, *tuberculous epididymitis* in males.

Although these complications often become apparent soon after the primary infection, there may be a *latent period* of years during which the tubercle bacilli remain dormant in their new site before initiating active disease.

Other sites of primary infection

The primary complex classically involves the lungs but the association of a primary focus with enlargement of the draining lymph nodes may be found elsewhere, e.g. the tonsils with cervical adenitis, the intestine with mesenteric adenitis. Tuberculosis may give rise to *'cold' abscesses*, i.e. without an acute inflammatory reaction. These abscesses, now rare, most often originated in a cervical lymph gland with discharge of caseous pus and the formation of draining sinuses. Psoas abscess is another example.

Post-primary infection

In post-primary tuberculosis the lesions, characteristically more localised and fibrotic, are modified due to the development of hypersensitivity by the host.

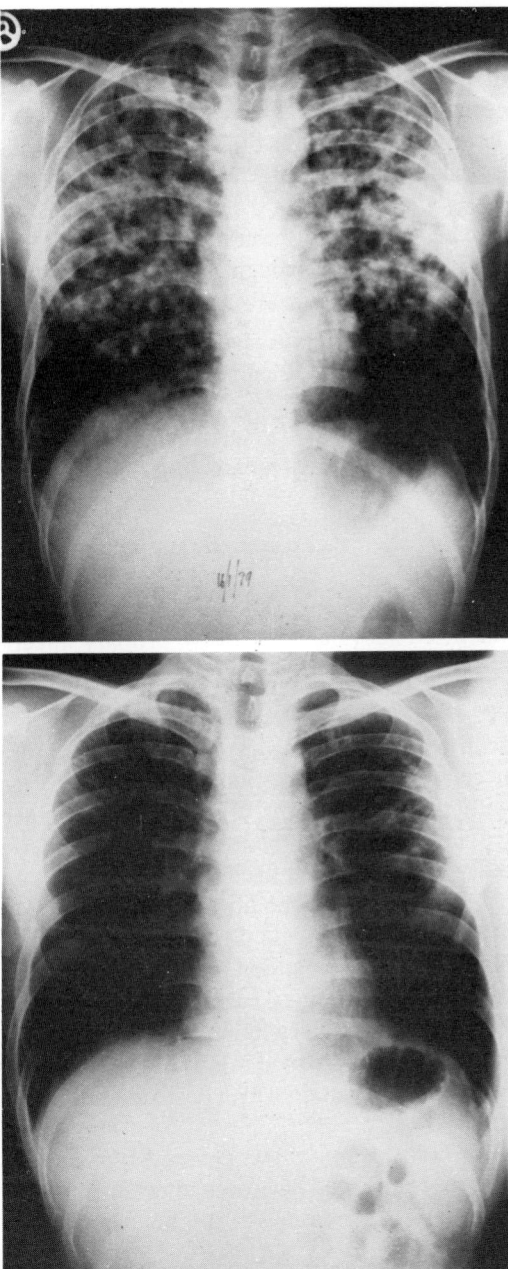

Fig. 38.1   Tuberculosis. (a) Chest x-ray of a patient with tuberculous bronchopneumonia. (b) Chest x-ray of the same patient 10 months after antituberculosis therapy. (Courtesy of Dr R. S. Kennedy.)

Post-primary tuberculosis usually involves the lungs with initial lesions in the apices: if untreated, chronic progressive disease may develop with areas of exudation and caseation surrounded by dense fibrosis; caseous lesions enlarge to coalesce and cavitate: cavities, sometimes large and containing fluid, can then be seen on X-ray: lymph node enlargement is much less marked in this form of tuberculosis.

Presenting symptoms are of non-specific ill health with fever: respiratory symptoms include cough, haemoptysis and a pneumonic illness that fails to respond to conventional antibiotics.

The source of post-primary infection is:

1. *Endogenous:* reactivation of latent foci formed at the time of primary infection.
2. *Exogenous:* reinfection by inhalation of infected respiratory secretions for a case of 'open' tuberculosis (i.e. with tubercle bacilli in the sputum).

Today in Britain patients are usually immigrants or the middle-aged and elderly.

Note. Post-primary infection may also disseminate and cause the manifestations listed above as progressive primary infection.

Delayed-type hypersensitivity in tuberculosis

Tubercle bacilli are readily phagocytosed but then multiply within mononuclear cells and resist digestion: this *intracellular parasitism* is associated with the development of delayed-type hypersensitivity (cell-mediated immunity) and of activated macrophages with an increased ability to kill ingested bacilli.

Delayed-type hypersensitivity modifies the host response to a second challenge with tubercle bacilli and differentiates primary from post-primary lesions. Koch recognised it a century ago when he demonstrated that inoculation of tubercle bacilli into a guinea pig already suffering from a primary lesion with caseating lymph nodes and disseminated infection resulted in only a small superficial lesion at the site of the second inoculation which healed rapidly without lymph node enlargement.

Skin tests in human infection: Delayed-type hypersensitivity develops in human beings who have been infected, most often without signs or symptoms of disease. It can be demonstrated by the tuberculin test.

Tuberculin test reagents:

1. Old tuberculin (OT) — a crude preparation of the filtrate of a culture of tubercle bacilli.

2. Purified Protein Derivative (PPD): obtained by chemical fractionation of OT; PPD is now the preferred material.

Potency of these preparations is assessed in international units and a standard dose used in tests.

Method: intradermal challenge either by

   a. *Injection* to raise a wheal (Mantoux test)

   b. *Multiple puncture* through a film of concentrated PPD applied to the skin (Heaf test).

Result: a positive test (i.e. indicating that delayed hypersensitivity has developed) produces induration surrounded by erythema at the site of injection after 24 to 48 h.

EPIDEMIOLOGY

Route of infection in the vast majority of cases is by inhalation of infected respiratory secretions coughed up by an actively infected person.

Reservoirs of infection are patients with 'open' phthisis, i.e. tuberculosis in which tubercle bacilli are coughed up in the sputum: there are still — even in Britain — many cases of unsuspected tuberculosis in the general population: elderly people living in poor housing conditions in cities are a particular problem.

Bovine tuberculosis was formerly common in cattle in Britain and resulted in some human cases due to drinking infected milk: the compulsory testing by tuberculin injection of all cattle and the slaughter of positive reactors has succeeded in the virtual eradication of the disease in cows: this measure combined with the *pasteurisation* of milk has resulted in human cases acquired from cattle becoming very rare.

Occupation: certain occupations were associated with a high mortality rate from tuberculosis: these are trades in which workers are exposed to the inhalation of stone or metal dust such as quarrymen, blasters and tin miners — especially if there is contamination with silica; doctors and nurses also show an increased risk of tuberculosis.

Age incidence: formerly largely a disease of childhood, tuberculosis is now most common in the elderly.

Race: Negroes and American Indians are more susceptible to tuberculosis than Caucasian (white) Americans; environmental or economic factors may contribute to this difference, which is probably also at least partly due to the severe effects of an infectious disease introduced into a community of people who were not exposed to it until relatively recently.

In the UK the disease is especially common in Asian immigrants.

Genetics: a tendency for tuberculosis to run in families is well recorded: to what extent this is due to an increased risk of exposure is hard to say; twin studies indicate that there may be some inherited increased susceptibility to infection.

Immunocompromised patients suffering from disease which affects the functioning of the immune system or on immunosuppressive therapy are extremely susceptible to tuberculosis: in them the infection may be rapid and overwhelming.

DIAGNOSIS

Specimen

These depend on the suspected site of infection.

Respiratory: sputum — if none available, laryngeal swab or washings from gastric lavage.

Meningitis: CSF.

Bone and joint: samples removed at operation or by aspiration.

Renal: early morning urine (i.e. the 'overnight' urine voided on waking).

Repeated specimens: three are usually necessary — especially in the case of suspected respiratory and renal tuberculosis.

Direct microscopy

Detection of typical acid-fast bacilli: in a smear stained by *Ziehl-Neelsen* method, confirms a diagnosis of tuberculosis in most cases. Care is necessary with urinary deposits, however, as these occasionally contain other types of acid and alcohol-fast bacilli.

Culture

All specimens must be cultured: culture is a more sensitive method of detection *Myco. tuberculosis* than direct demonstration and even if acid-fast bacilli have been detected, culture is necessary for antibiotic sensitivity testing.

Specimens usually require some treatment before inoculation of culture medium such as *Lowenstein-Jensen medium.*

Sputum: is 'concentrated' by treatment with sodium hydroxide to liquify the sputum and kill contaminating bacteria; after centrifugation and neutralization with hydrochloric acid, the deposit is cultured.

CSF: is centrifuged before culture.

Bone and joint samples: are disrupted in a sterile grinder before culture.

Urine: is centrifuged and 'concentrated' as for sputum before culture.

Cultures: are examined weekly: *Myco tuberculosis* and *Myco bovis* can be distinguished by special laboratory tests including cultural characteristics.

Colonial morphology:
1. *Human strains* grow as dry, crumbly yellowish colonies — often described as 'rough, tough and buff'.
2. *Bovine strains* grow as smoother colonies; glycerol does not enhance their growth in contrast to human strains.
   Cultures should not be discarded as negative until they have been incubated for 8 weeks.

Guinea pig inoculation. Guinea pigs are very susceptible to tuberculosis and before adequate culture media had been developed, they were extensively used in diagnosis. Animals injected intramuscularly with clinical material containing tubercle bacilli develop disseminated disease in 4 to 8 weeks. Rarely require to be used nowadays.

TREATMENT

The discovery of streptomycin in the 1950s revolutionised the treatment of tuberculosis. Resistant strains, however, emerge readily and combinations of two, or more usually three, drugs are used to minimise the risk of this.

Triple therapy: with streptomycin, isoniazid and para-aminosalicytic acid (PAS) was widely and successfully used for treating tuberculosis; it has now been replaced with other regimens.

Streptomycin has to be given by injection: it is also toxic and side-effects include deafness and vertigo due to damage to the eighth nerve and nephrotoxicity.

PAS is taken orally but the tablets are large and unpalatable: nausea and sickness are fairly common complications.

Present recommended triple therapy is a combination of:
Isoniazid
Rifampicin
Ethambutol

These drugs (given orally) are less toxic and better tolerated by patients than the earlier triple therapy. The duration of treatment is now much shorter — 6 months if there is no cavitation and 9 months if this is present; ethambutol is stopped after 8 weeks; 3-month courses are under trial.

Alternative or second-line antituberculous drugs:
Streptomycin   thiacetazone
Pyrazinamide   cycloserine
Prothionamide   capreomycin

The choice of drugs should be reviewed when the sensitivity of the infecting strain to a range of antituberculous drugs has been reported. In patients successfully treated, cultures from the site of infection will soon become negative for tubercle bacilli. Should they remain positive, the sensitivity of the isolated organism must be tested again to determine if drug resistance has emerged.

PREVENTION AND CONTROL

1. *Socio-economic factors*
Tuberculosis is a disease associated with poverty, malnutrition and overcrowding. Improvement in living standards reduces the incidence: thus deaths from respiratory tuberculosis in the UK fell by two-thirds between 1910 and 1950, i.e. before the introduction of immunisation and chemotherapy.

2. *Identification of infectious cases*
Patients with 'open' phthisis are the important source of infection especially in overcrowded conditions; when diagnosed — either by clinical illness or during 'case finding' campaigns — they should be isolated and treated without delay; effective chemotherapy renders nearly all cases non-infectious within a few weeks.

3. *Immunisation*
Immunisation is with live bovine tubercle bacilli attenuated by growth in bile-containing medium. The vaccine is known as BCG (Bacille Calmette Gúerin) after the workers who developed it.

4. *Chemoprophylaxis*
Recommended for those at exceptional risk, e.g. children in contact with a known open case or an elderly relative. The drugs chosen are usually isoniazid and rifampicin administered for 6 months.

LEPROSY

The scourge of the ancient world and still afflicting millions of patients mainly in China and India today. Despite the introduction of effective treatment, many, possibly the majority of cases, remain untreated.

CLINICAL FEATURES

Leprosy is a slow, chronic and progressive infection which mainly affects the skin: skin lesions present as nodules or thickened patches and are often associated with thickening of the peripheral nerves and anaesthesia; there may be muscle weakness and ulceration and trophic changes in tissues of the extremities leading to the distressing mutilation characteristic of the disease. The causal organism multiplies only at low temperature (30°C) and so lesions mainly affect the skin and exposed cooler extremities such as nose and ears.

Causal organism: Myco. leprae.

Incubation period is long, usually from 3 to 5 years.

Two forms of leprosy are recognised:

Lepromatous: in which lesions are diffuse and scattered and may involve mucous membranes, e.g. of the nose: numerous *Myco. leprae* are present in the lesions; this form of the disease is progressive and severe (Fig. 32.2).

Tuberculoid: in which the lesions are localised but show early nerve involvement with anaesthesia: there are only scanty *Myco. leprae* in the

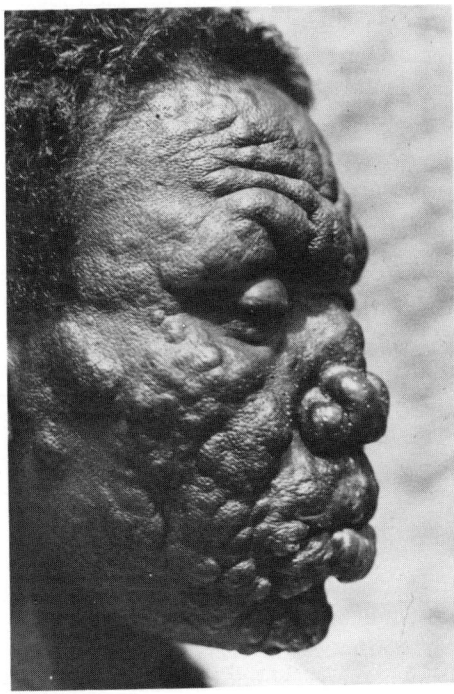

Fig. 38.2   The facial disfiguration of leprosy. This patient has the leonine face sometimes seen in lepromatous leprosy. (Photograph by Dr R. St. C. Barnetson.)

lesions: in this type of leprosy, lesions tend to be benign and the disease is often self-healing.

Delayed-type hypersensitivity plays the major part in determining the response of the host to the infection and therefore the clinical type of disease produced: patients with a high degree of hypersensitivity develop the tuberculoid form; in those with no or only a low degree of cell-mediated immunity, the disease tends to be lepromatous.

Lepromin test — the intradermal inoculation of an extract of *Myco. leprae* can be used to demonstrate cell-mediated immunity: a red papule develops at the site of injection in about 48 hours. Positive in tuberculoid leprosy: usually negative in lepromatous leprosy.

EPIDEMIOLOGY

Spread in the community is slow. Leprosy is of *very low infectivity* and even now its mode of spread is uncertain: prolonged contact with cases is probably necessary and the route of infection may be contamination of small cuts and abrasions in the skin; patients with lepromatous leprosy of the nose or skin are probably the main source of infection.

Geographic distribution is widespread (10 million cases worldwide), especially in tropical climates: the disease, however, is found in the southern United States and in some Mediterranean countries in Europe.

DIAGNOSIS

Specimens: skin biopsy or scrapings from lesions in the nasal mucosa.

Direct demonstration of acid-fast bacilli in smears or sections from lesions.

Myco. leprae does not grow on artificial media in the laboratory: it can be cultivated *in vivo* by inoculation of the food pads of mice and armadillos.

TREATMENT

Dapsone: 4-4—diacetyl-diaminodiphenyl sulphone (DDS) is effective in treating the disease; the drug is unpleasant to take causing side effects such as anaemia, liver dysfunction and skin rashes; treatment must be continued for years and often for life.

Rifampicin is proving of value but resistance can be a problem and the disease may recur when treatment is stopped.

Clofazimine.

CONTROL

The traditional isolation of patients prevents early cases getting treatment; hospitalisation may be indicated at the start of treatment but thereafter patients regarded as non-infectious can resume their normal activities.

Vaccine: BCG vaccine may offer a degree of protection especially in children.

DISEASES DUE TO ATYPICAL 'ANONYMOUS' MYCOBACTERIA

Atypical mycobacteria are a rather ill-defined group of mycobacteria with growth characteristics, biochemical activity and animal pathogenicity which differ from those of *Myco. tuberculosis:* for example, many produce pigment on culture and grow at lower or sometimes higher temperature; the pathogenicity and infectivity of these organisms is much lower and in many instances they are present in lung infections as 'fellow travellers' along with *Myco. tuberculosis* and do not play a significant pathogenic role.

Three main diseases due to atypical mycobacteria are recognised:

1. *Pulmonary infection:* with clinical features resembling pulmonary tuberculosis but generally a milder, slower and more chronic disease; due to *Myco. kansasii, Myco. avium/intracellulare* and *Myco. xenopi.*
2. *Cervical adenitis:* swelling and purulent discharge from the cervical glands in children; most often due to *Myco. scrofulaceum.*
3. *Skin ulcers:* contracted most often in swimming pools due to contamination of skin cuts and abrasions: the ulcers are benign and heal spontaneously although possibly not for a fairly prolonged period, e.g. 1–2 years; due to *Myco. marinum* and *Myco. ulcerans;* these atypical mycobacteria grow best at 30–33°C.

TREATMENT

Many atypical mycobacteria, although sensitive to rifampicin, are characteristically resistant to other antituberculous drugs, particularly streptomycin and isoniazid. Drug combinations for treatment have to be chosen on the results of laboratory sensitivity tests.

39

Infections of bone and joint

Predominantly infections of childhood: delay in diagnosis and inadequate treatment may result in protracted illness followed by permanent disability.

BONE INFECTIONS

Acute osteomyelitis

Clinical features: most common in children under 10 years old; distal femur, proximal tibia and proximal humerus the classical sites. Presentation is with fever, bone pain and local tenderness: the child is reluctant to move the affected limb. There may be a history of preceding mild trauma to the involved bone.

Causal organisms: Staphylococcus aureus (75 per cent or more of cases); other organisms include *Haemophilus influenzae, Streptococcus pneumoniae, Strep. pyogenes.*

Source: not always apparent; sometimes a septic focus elsewhere, e.g. a boil. Spread to bone is haematogenous. Infection at all ages may follow trauma (e.g. compound fracture) or surgical operation.

LABORATORY DIAGNOSIS

Bacteriological diagnosis and antibiotic sensitivity tests are essential for successful management.

1. *Blood culture:* positive in a high proportion of cases; a series of cultures may be necessary.
2. *Culture* of pus collected from the diseased bone either by needle aspiration or at open operation.

TREATMENT

Antibiotics alone are often effective, especially if started early: surgery is necessary if there is evidence of bone destruction or accumulation of pus.

Antibiotic treatment: initially parenteral, later, the oral route is adequate; in the absence of positive cultures, give flucloxacillin on the assumption that the pathogen is *Staph. aureus.* Alternative drugs include erythromycin, fusidic acid and clindamycin — (the latter two penetrate bone). If a penicillin-sensitive *Staph. aureus* is later isolated penicillin should be substituted for flucloxacillin. Sometimes a combination of antistaphylococcal drugs is preferred.

Strep. pneumoniae: penicillin

Strep. pyogenes: penicillin

H. influenzae: ampicillin

Note. Treatment must be continued for several weeks.

Chronic osteomyelitis

Sometimes due to progression from acute osteomyelitis but may present as a primary infection of bone.

Clinical features: pain, bone destruction with the formation of sequestra and discharging sinuses. If the vertebral column is involved there may be vertebral collapse resulting in paraplegia.

Causal organisms: Staph. aureus — the most common pathogen; rarely *Mycobacterium tuberculosis, Salmonella typhi* (and other salmonellae), *Brucella species.*

LABORATORY DIAGNOSIS

Isolation of the causal agent from:

  (i) A discharging sinus or material obtained at operation.
 (ii) Blood or bone marrow culture.
(iii) Possibly from an infective focus elsewhere (e.g. in tuberculosis).

TREATMENT

Surgery is usually necessary for the drainage of pus and for the removal of sequestra.

Antibiotics:

1. *Chronic staphylococcal osteomyelitis:* long-term flucloxacillin, fusidic acid or clindamycin; alone or in combination.
2. *Tuberculous osteomyelitis:* isoniazid, ethambutol and rifampicin for 2 years.
3. *Salmonella osteomyelitis:* chloramphenicol or cotrimoxazole.
4. *Brucella osteomyelitis:* tetracycline or cotrimoxazole.

JOINT INFECTIONS

Infective arthritis

Clinical features: in general, only one joint is involved. Acute infective arthritis is characterised by sudden onset of fever and pain. There is swelling with limitation of movement in the joint. Crippling sequelae are common despite antibiotic therapy.

Migratory polyarthralgia and fever are features of rheumatic fever and this diagnosis should be considered if there is a history of recent sore throat.

Causal organisms:
  Staph. aureus
  Strep. pneumoniae and other streptococci
  H. influenzae
  Neisseria gonorrhoeae
  N. meningitidis
  Brucella species
  Salmonella species
  Non-sporing anaerobes, e.g. *Bacteroides species*
  Myco. tuberculosis

Source: haematogenous spread from a remote infective focus is the rule but infection may encroach upon the joint from adjacent osteomyelitis.

LABORATORY DIAGNOSIS

1. Blood culture or, where indicated, marrow culture.
2. Examination of fluid aspirated from the joint
   a. *Direct film:* observe for polymorphs and bacteria; the appearance may allow a presumptive diagnosis and advice regarding immediate chemotherapy
   b. *Culture:* on a variety of media to isolate the causal pathogen.
3. Culture of specimens from any other infected site, e.g. throat, genital tract, meninges. If tuberculosis is suspected, examine sputum and urine.
4. Serological tests for salmonellosis and brucellosis.

TREATMENT

Antibiotic therapy should be started as soon as diagnostic specimens have been taken on a 'best-guess' basis, paying attention to clues in the history or on clinical examination which suggest the probable infective agent. When a causal organism has been isolated, the drug of choice is the same as in osteomyelitis. For arthritis due to *N. gonorrhoeae* or *N. meningitidis*, give penicillin.

Infection in prosthetic joints

The insertion of artificial joints to replace those damaged by disease is now a routine orthopaedic operation. The success rate is high (96 per cent) but failure when it occurs is usually due to infection. The prostheses are made of metal and plastic and the introduction of inert material predisposes to infection, most often with organisms of low pathogenicity.

Clinical features: as a rule infection presents soon after operation but may be delayed for several months. The new joint becomes painful and movement is limited.

Causal organisms: although *Staph. aureus* and other recognised pathogens may be responsible, the usual infecting organisms are skin commensals — *Staph. epidermidis* and corynebacteria.

Source: direct contamination of the site at the time of operation with bacteria.

(i) From the patient's skin
(ii) Dispersed by the operating team
(iii) Present in the theatre air

Late infection may be due to organisms settling in the implant following a transient asymptomatic bacteraemia.

Prevention: careful preparation of the skin site and scrupulous surgical technique: operations may be carried out either in a specially ventilated theatre or within a plastic isolator tent.

Peroperative antibiotic prophylaxis e.g. with flucloxacillin, cephalosporins or fusidic acid.

Laboratory diagnosis of established infection may be impossible because specimens are not always available. Bacteria in the exudate of the superficial surgical wound are not necessarily the same as those present deep at the site of infection.

Treatment: often blind, with flucloxacillin, cephalosporins, fusidic acid; gentamicin may be given in addition.

40

Sexually transmitted diseases

Several infectious diseases are transmitted predominantly or entirely by sexual intercourse. Often the causal organisms are delicate and do not survive long outside the body. Sexually transmitted diseases naturally tend to produce genital lesions but several can give rise to systemic, sometimes severe, disease.

Sexually transmitted diseases affect homosexual partners as well as heterosexual relationships. Variations in sexual behaviour can result in these diseases producing lesions in rectum or oropharynx.

The stigma of attending a 'VD clinic' is now less than in former days: increasingly patients with sexually transmitted disease are seen at 'urogenital clinics'; in fact many of the patients are suffering from non-sexually acquired infection and their disease is not evidence of marital infidelity or promiscuity.

The main sexually transmitted diseases are listed in Table 40.1.

Gonorrhoea

A worldwide disease which has been steadily increasing over the past 20 years: possibly due to changing social attitudes, increased travel amongst the young, urbanisation, etc.

Causal organism: Neisseria gonorrhoeae.

CLINICAL FEATURES

Acute onset of purulent urethral or vaginal discharge: often with some dysuria and frequency in males; asymptomatic cases are common in females. The disease sometimes involves the rectum or the oropharynx.

Complications: due to local spread — prostatitis, epididymitis in males — rarely urethral stricture in untreated cases, salpingitis in women; occasionally septicaemia, arthritis, meningitis due to haematogenous spread.

Ophthalmia neonatorum or gonococcal conjunctivitis in the newborn infected during birth from maternal disease.

Table 40.1 Sexually transmitted diseases

Disease	Cause
Gonorrhoea	*Neisseria gonorrhoeae* (the gonococcus)
Non-specific urethritis	*Chlamydia trachomatis* ? Unknown
Trichomoniasis	*Trichomonas vaginalis*
Vaginal thrush*	*Candida albicans*
Syphilis	*Treponema pallidum*
Vaginitis	*Haemophilus vaginalis*
Genital herpes	Herpes simplex virus type 2
Genital warts	Papilloma virus
Hepatitis B*	Hepatitis B virus
Chancroid	*Haemophilus ducreyi*
Lymphogranuloma venereum	*Chlamydia trachomatis*
Granuloma inguinale	*Donovania granulomatis* (a Klebsiella-like microorganism)
Pubic lice (crabs)	*Pthirus pubis*
Genital scabies	*Sarcoptes scabei*

*Not always sexually transmitted.
Note: Only gonorrhoea, non-specific urethritis, trichomoniasis, thrush, syphilis and
H. vaginalis will be dealt with in this chapter

Gonococcal vulvovaginitis: a rare form of gonorrhoea in young girls either following a sexual offence or sometimes acquired innocently by contact with infected exudates or fomites.

DIAGNOSIS

Specimens: urethral, cervical smears and swabs, swabs directly plated or transported to laboratory in Stuart's or Amies' transport medium.

Direct Gram film: examine for typical intracellular Gram-negative diplococci in smears (Fig. 40.1): often convincingly positive in males but, due to difficulty in interpreting the microscopic appearances in the mixed normal flora, less useful in females.

Culture: on modified Thayer-Martin medium which is a nutrient lysed–blood agar made selective for *N. gonorrhoeae* by the addition of antibiotics — vancomycin, colistin, nystatin and trimethoprim; chocolate agar can be used but the isolation rate is lower. Thayer-Martin medium is particularly useful in the culture of specimens from sites heavily contaminated with other organisms, e.g. vagina or rectum: incubate in CO_2 for 48 h.

Observe: for typical translucent colonies reacting as oxidase-positive by turning purple on addition of tetramethyl-*p*-phenylene-diamine.

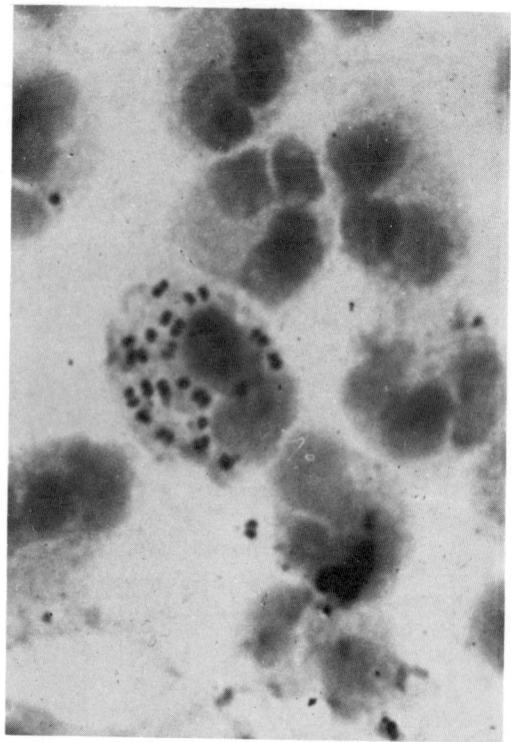

Fig. 40.1   Gonorrhoea. Urethral pus with typical intracellular diplococci (magnified × 4000).

Identify: by acid production (oxidatively) from glucose but not lactose, maltose or sucrose.

TREATMENT

Patients tend to default and whenever possible antibiotics should be given in one curative dose. Numerous different regimens have been proposed.

Standard treatments include;

1. A large intramuscular dose of penicillin or oral ampicillin along with oral probenecid to delay renal excretion.
2. Doxycycline (a tetracycline) in a large oral dose.

Penicillin resistance: normally *N. gonorrhoeae* is extremely sensitive to penicillin but two types of resistance have now been encountered.

   (i) *Low-level resistance:* first recognised in 1958: strains are still sensitive to low concentrations of penicillin; relatively common; not due to β-lactamase production.

(ii) *High-level resistance:* first recognised in 1976; strains totally insusceptible to penicillin; uncommon in the UK (about 200 isolates identified in 1980); due to a plasmid coded β-lactamase.

There are two types of β-lactamase producing strain; one originated from West Africa, the other from South-East Asia. Resistance in African strains is due to a small plasmid which cannot be transmitted to other gonococci. However, Asian organisms contain, in addition, a second larger plasmid which can *mobilise* the smaller and mediate transfer to other previously sensitive strains; this has serious epidemiological implications.

Penicillin-resistant gonorrhoea should be treated with cefuroxime or spectinomycin.

Non-specific urethritis or cervicitis
Non-specific genital infection is now the commonest sexually transmitted disease in Britain. Predominantly seen in males — presumably because infection in females is often symptomless.

Causal organism(s): probably due to more than one infectious agent: in about half the cases, infection with *Chlamydia trachomatis* can be demonstrated.

CLINICAL FEATURES

Virtually indistinguishable from gonorrhoea — i.e. acute purulent urethral discharge — but *N. gonorrhoeae* cannot be demonstrated nor isolated: sometimes cervicitis, but usually symptomless, in females.

Reiter's disease or syndrome is a triad of urethritis, arthritis and conjunctivitis (with or without iritis): often seen as a complication of non-specific urethritis; sometimes only two of the three symptoms are present.

DIAGNOSIS

Mainly on the basis of clinical signs and symptoms in the presence of negative gonococcal cultures, but attempts should be made to look for chlamydial infection.

Specimens: smears and swabs of urethral or cervical discharge; the latter transported in special transport medium.

Smears: examine by indirect immunofluorescence using specific antiserum.

Culture: in McCoy tissue culture cells treated with cycloheximide or irradiation.

Observe: for intracytoplasmic inclusions by Giemsa stain.

TREATMENT

Tetracycline for 7–10 days — but relapses are common.

Trichomoniasis

A common disease in women: mainly if not entirely transmitted as a result of sexual intercourse with males who have a symptomless infection.

Causal organism: a pear-shaped protozoan — *Trichomonas vaginalis* motile by four flagella (Fig. 4.2).

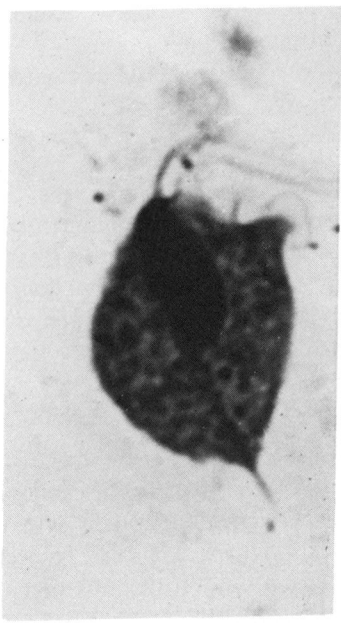

Fig. 40.2 *Trichomonas vaginalis.* This common protozoan is pear-shaped with four flagella (magnified × 250).

CLINICAL FEATURES

Vaginal discharge, typically frothy, offensive and usually greenish yellow: there may also be urethral discharge and excoriation of vulva and perineum; the mucosa of vagina and cervix is reddened and inflamed. The bladder may become involved causing dysuria and frequency. Infection is usually symptomatic but varies from an acute severe vaginitis to a mild low-grade or even symptomless infection.

DIAGNOSIS

Specimen swab of vaginal discharge in Stewart's or Amies' transport medium.

Direct wet film: observe for motile protozoa.

Culture: in Fineberg's medium for 5 days: examine for motile protozoa at 2 and 5 days.

TREATMENT

Metronidazole: orally for 7 days or a large single dose. Whenever possible the male partner should be identified, examined and treated.

Thrush (candidosis)

Vaginal thrush is another condition seen in females; genital candidosis is rare in men.

Causal organism: Candida albicans, a normal commensal of mucous membranes including the vagina and bowel. Infection may be endogenous precipitated by systemic disease (e.g. diabetes) or drug treatment (e.g. broad-spectrum antibiotics); but certainly a proportion of cases are sexually transmitted.

CLINICAL FEATURES

White, membranous patches with itching and irritation of vulva or vagina: white, thick or sometimes watery discharge may be present; many cases are virtually symptomless.

DIAGNOSIS

Can usually be made clinically but can be confirmed in the laboratory.

Specimen: swab.

Direct Gram film: look for characteristic Gram-positive yeasts with pseudohyphae.

Culture: on Sabouraud's medium.

Observe: for typical waxy-surfaced colonies.

Identify: by observation of germ tubes on growth in serum.

TREATMENT

Topical application of Nystatin or nitromidazoles (e.g. miconazole).

Syphilis

Now a rare disease but still important because of its severity and long-term effects if inadequately treated or missed.

Causal organism: a spirochaete — *Treponema pallidum.*

CLINICAL FEATURES

A disease which is seen in four clinical stages.

Incubation period: 2–4 weeks.

Primary: chancre or papule usually on the genital area: this ulcerates but heals spontaneously in 3 to 8 weeks; there is painless enlargement of local lymph nodes.

Secondary: 6 to 8 weeks later: rash; the lesions may coalesce in intertrigenous areas (especially perianal region) to form wart-like condylomata lata. There is generalised lymphadenopathy in half the patients at this stage and snail-track mucosal ulcers in the mouth in about a third.

Tertiary: 3 to 10 years after the primary lesion: *gumma* or granulomatous lesions in skin, mucous membrane or bones.

Late or quaternary: 10 to 20 years after primary syphilis: two main clinical forms:

1. *Cardiovascular:* aortitis, aneurysm, aortic incompetence, coronary ostial stenosis.
2. *Neurosyphilis:* tabes dorsalis, general paralysis of the insane, meningovascular syphilis.

Latent syphilis: the disease may lie dormant for many years with no clinical symptoms (but with positive serology). Latent syphilis may develop into cardiovascular or neurosyphilis.

Congenital syphilis: infection in the mother may be transmitted placentally to the fetus: affected babies have domed skulls, deformed teeth, rash and jaundice.

DIAGNOSIS

Direct demonstration of spirochaete in fluid or scrapings from chancre or ulcerated secondary lesions by dark-ground microscopy.

Serology. There are three main tests (Table 40.2) — two are specific (TPHA and FTA/Abs) for *Tr. pallidum* but one — the VDRL — detects reagin or antibody which reacts with non-treponemal lipoidal antigen: it probably detects antibody formed as a result of tissue damage and the titre gives an indication of disease activity.

Note: The CSF in neurosyphilis gives positive results in these tests.

Note. Biological false-positive reactions are not uncommon with the VDRL test especially in pregnancy, febrile states, malaria.

Wassermann reaction: the classical serological test for syphilis; no longer in routine use: it is a complement fixation test using lipoidal antigen to detect reaginic antibody.

Table 40.2   Serological tests for syphilis

Stage of disease	VD Reference Laboratory (VDRL)	*Treponema pallidum* haemagglutination (TPHA)	Fluorescent Treponema antibody absorbed (FTA (Abs))
Primary	+ or −	−	+
Late primary	+	+ or −	+
Secondary and tertiary	+	+	+
Late (quaternary)	+	+	+
Latent	+ or −	+	+
Treated syphilis	−	+	+

Note. The success of treatment can be monitored by the VDRL test which becomes negative on successful treatment. The other two tests remain positive.

Reiter protein complement fixation test detects group specific antibody to the related Reiter treponeme.

Treponema pallidum immobilisation test (TPI): measures inhibition of motility of a suspension of live *Trep. pallidum*; difficult to carry out and now almost completely replaced by the FTA Abs test.

Congenital syphilis: may be difficult to diagnose if the mother's serum is positive because IgG of maternal origin crosses the placenta. Demonstration in the baby's blood of specific IgM by TPHA and FTA(Abs) indicates infection of the fetus.

TREATMENT

Penicillin:

Primary: large doses continued for 15 to 21 days.

Late or latent: large doses for 21 days, usually followed by 10 injections at weekly intervals.

Erythromycin or tetracycline can be used if the patient is hypersensitive to penicillin; cephaloridine may be effective but some patients show cross-reacting hypersensitivity.

Haemophilus vaginalis vaginitis

Many episodes of vaginitis remain unexplained. One possible causal agent, *Haemophilus vaginalis,* is currently the subject of interest and debate. Infection is probably transmitted sexually although infectivity appears to be low: the bacterium can be isolated from male sexual partners.

Causal organism: Haemophilus vaginalis (previously *Corynebacterium vaginale*) a Gram-variable microaerophilic bacillus of uncertain classification.

Clinical features
Grey-white thin offensive vaginal discharge.

Diagnosis
Microscopy: demonstration of 'clue cells', i.e. squamous epithelial cells with many adherent bacilli.

Isolation of a slow-growing fastidious bacillus on suitable culture media after 48 h incubation.

Treatment
Metronidazole produces clinical cure — unusual for an infection due to a microaerophilic (and not an obligate anaerobic) microorganism.

CONTROL OF SEXUALLY TRANSMITTED DISEASES

Clearly it is not easy to trace partners of patients with sexually transmitted diseases. In fact the steady increase in the incidence of these diseases in the past two decades illustrates the impossibility of doing so. Nevertheless determined attempts should be made to persuade patients to name consorts and for the consorts to submit themselves to examination and treatment.

41

Infections of the eye

EYELID INFECTIONS

Blepharitis
Red eyelids with fine crusts at the roots of the lashes; may progress to ulceration with destruction of the lids.

Stye
An abscess or small boil in one of the glands of the lash follicles; external styes point to the outer side of the lid margin, internal styes to the inner surface of the eyelid.
Causal organism: Staphylococcus aureus.
Source: endogenous, e.g. from the anterior nares or implantation via the fingers from a septic lesion elsewhere.
Diagnosis: culture of pus or a swab: antibiotic sensitivity tests to topical drugs.
Antibiotic treatment often unnecessary.

CONJUNCTIVITIS

The conjunctival sac is normally colonised by *Staphylococcus epidermidis* and corynebacteria (notably *Corynebacterium xerosis*). Tears act as a defence mechanism by flushing away foreign material; in addition, they contain lysozyme, an enzyme which degrades the mucopeptide of the bacterial cell wall especially that of Gram-positive organisms.
Conjunctivitis is the result of the invasion of the conjunctival sac by pathogens; children and the elderly are most often affected.
Clinical features: redness of the eye, swollen lids: usually bilateral; the eyes feel hot, gritty and sticky: excessive lacrimation with a mucoid or purulent discharge.
Causal organisms: Staph. aureus, Haemophilus influenzae, Streptococcus pneumoniae, Moraxella lacunata ('angular' conjunctivitis).

Source and spread: sometimes from a stye: often the result of cross-infection via contaminated fingers or fomites. Epidemics, especially in institutions, are common.

Note. Conjunctivitis is often due to adenoviruses.

KERATITIS

Keratitis is inflammation of the cornea.

Clinical features: pain, photophobia, lacrimation and lid spasm: circumcorneal injection: ulceration with grey-white areas and loss of corneal reflex. Severe ulceration can be accompanied by iritis and collection of pus at the lower recess of the anterior chamber.

Cause: minor trauma, e.g. a foreign body with secondary infection and as a complication of conjunctivitis.

Note. Recurrent dendritic ulcer is a serious form of keratitis due to herpes simplex virus.

OPHTHALMIA NEONATORUM

Infection of the baby's eyes acquired either during birth from the maternal genital tract or from some external source after delivery.

Causal organisms:

1. *From the genital tract*
   a. *Neisseria gonorrhoeae* — *gonococcal ophthalmia:* now rare but formerly an important cause of blindness in children. Clinically, severe purulent conjunctivitis with a tendency to involve the cornea causing ulceration and even perforation. Usually develops within 36–48 h of birth. Can be prevented by instillation of 1 per cent silver nitrate solution into the eyes at birth.
   b. *Chlamydia trachomatis* — *inclusion conjunctivitis:* clinically, an acute conjunctivitis with mucopurulent exudate: less severe than gonococcal ophthalmia and with a longer incubation period (5-12 days).
   c. *Streptococci of Lancefield group B.*

2. *From an external source*
   Staphylococcus aureus — *'sticky eye':* the commonest eye infection, probably spread via medical and nursing staff. Outbreaks of staphylococcal sticky eye in maternity-hospital nurseries are well recognised.

DIAGNOSIS

Bacterial infections

Specimen:
1. Exudate collected directly from the patient's eye with a platinum bacteriological loop: films and cultures should if possible be made at the bedside.
2. Swab of exudate placed in Stuart's transport medium.

   Direct film: examine by Gram's stain for characteristic bacteria.

   Culture: on blood and chocolate agar plates incubated for 48 h at 37°C in the presence of 5-10 per cent carbon dioxide.

   Observe: for growth and identify by usual methods.

Chlamydial infections

Specimen: conjunctival scrapings.

   Direct film: examine by immunofluorescence.

   Culture: in irradiated or cycloheximide-treated McCoy cells for 2–3 days.

   Observe: for intracytoplasmic inclusions by Giemsa stain.

ORBITAL CELLULITIS

A serious infection of the cellular tissues of the orbit which may follow penetrating injury or spread of infection from the nasal sinuses. Painful swelling and protrusion of the eyeball are accompanied by general systemic upset. Complications include panophthalmitis, meningitis, brain abscess and cavernous sinus thrombosis.

Causal bacteria: Staph. aureus, other pyogenic bacteria.

Surgical incision may yield pus — an essential specimen for accurate bacteriological diagnosis.

PANOPHTHALMITIS

Inflammation of the whole substance of the eye which may result in total destruction. Usually due to:
1. Spread of infection from a perforated corneal ulcer or orbital cellulitis.
2. Penetrating wounds — traumatic or surgical.

   Causal bacteria:
   a. *Staph. aureus*
   b. *Pseudomonas aeruginosa, Klebsiella species, Proteus species* and *Escherichia coli*

Diagnosis: culture of a swab taken from the surface of the eye may give misleading results: as soon as possible a sample of pus should be aspirated from the anterior chamber of the eye.

TREATMENT

Superficial infections
These usually respond readily to local treatment

1. *General measures:* removal of exudate, irrigation with simple solution, e.g. saline.
2. *Antibiotics:* either as drops or ointments — chloramphenicol, sulphacetamide, penicillin, aminoglycosides (e.g. framycetin, streptomycin, neomycin), tetracycline, polymyxin and bacitracin. The most widely prescribed are chloramphenicol and framycetin.

Chlamydial infections should be treated orally by tetracycline or erythromycin.

Deep infections
More aggressive antibiotic treatment is necessary:

1. *Systemic antibiotics* unfortunately, penetrate the eye poorly. Concentrations attained in the aqueous exceed those in the vitreous.
2. *Subconjunctival injection:* the route of choice: although painful, up to 1 ml of solution can be injected — penicillin, gentamicin, neomycin, polymyxin can be given in this way.

CHOROIDITIS AND CHORIORETINITIS

Granulomatous reaction in the choroid or retina is usually the result of invasion by microorganisms or parasites: clinically, grey-white choroidoretinal lesions which heal to form scars.

Causal organisms:

1. *Bacteria:* Mycobacterium tuberculosis, Brucella abortus, Treponema pallidum (now very rare).
2. *Viruses:* rubella and cytomegalovirus infection — either congenital or as a rare complication of renal transplantation due to immunosuppression.
3. *Protozoa:* Toxoplasma gondii (in congenital and acquired toxoplasmosis).
4. *Helminths:* Toxocara canis (mainly retinitis).

42

Oral and dental infections

There are four main types of oral and dental infections:
1. *Non-specific localised infections* caused by oral commensal bacteria usually in mixed culture, e.g. dental caries, periodontal disease and periapical abscess.
2. *Specific localised infections* caused by oral commensal microorganisms, e.g. actinomycosis and candidosis.
3. *Specific systemic infections with oral symptoms,* e.g. tuberculosis, syphilis, gonorrhoea and various viral infections (e.g. herpes simplex, mumps, measles and Coxsackie group A viruses).
4. *Systemic infections without oral symptoms but often caused by oral commensal bacteria,* e.g. infective endocarditis

1. NON-SPECIFIC LOCALISED INFECTIONS

DENTAL CARIES AND PERIODONTAL DISEASE

Dental caries and periodontal disease are the most common of human infections. Generally they are not directly attributable to a specific pathogenic agent but are due to the activity of a mixture of commensal oral bacteria in the form of dental plaque. A diagram of a tooth and its related structures are shown in Fig. 42.1.

Dental plaque
Composition: dental plaque consists of bacteria (2.5×10^{11} cells per g wet weight) embedded in an organic matrix derived partly from salivary glycoproteins, the gingival exudate and microbial extracellular polymers: it has a heterogeneous composition which is affected by diet and the stage of its development.

There are three main sites where dental plaque is deposited:

a. Supragingival (above the gum margin)
b. Subgingival (below the gum margin) usually with established periodontal disease
c. Occlusal (in the fissures of the teeth)

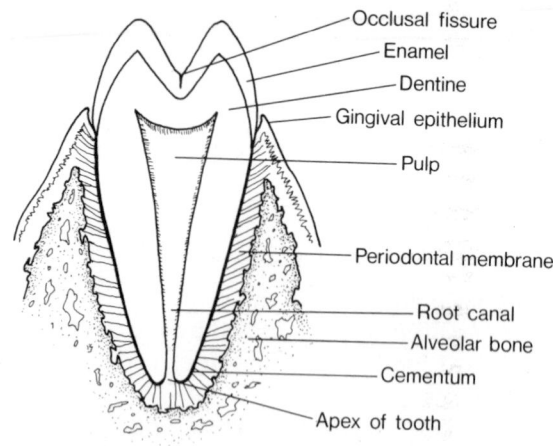

Fig. 42.1 Diagram of a tooth and its related structures.

The microbial composition of supragingival plaque is shown in Table 42.1. Subgingival plaque has a similar composition, but contains more anaerobic Gram-negative bacilli (*Bacteroides species*), spirochaetes and campylobacter. Occlusal plaque consists mainly of streptococci (*Strep. sanguis, Strep. salivarius* and *Strep. mutans*) with smaller numbers of veillonella, lactobacilli and *Haemophilus species*.

Table 42.1 Bacterial composition of 'mature' supragingival dental plaque

	Mean per cent viable count
Streptococci	27
Gram-positive bacilli (mainly Actinomyces)	41
Neisseriae	2
Veillonella	13
Bacteroides species	10
Fusobacteria	3
Campylobacter species	1
Others:	3
Spirochaetes	*Not found*
Individual species:	
Streptococcus mutans	2.0
Streptococcus sanguis	6.0
Streptococcus salivarius	0.5
Actinomyces israelii	17
Actinomyces viscosus } *Actinomyces naeslundii* }	19

Development of supragingival plaque
If a tooth surface is thoroughly cleaned, plaque develops as follows:

(i) Within minutes a thin layer of salivary glycoprotein is deposited (acquired pellicle).

(ii) Within hours, 10^6 viable bacteria per mm^2 of tooth surface can be isolated, mainly *Strep. sanguis*, *Strep. mitis* and *Neisseria species*. These and other bacteria become attached to the tooth surface and to one another by means of extracellular polymers.

(iii) Plaque development is affected by microbial interactions which affect microbial adhesion and metabolic activity, e.g. *Veillonella species* can use lactate produced by other bacteria from sucrose and so reduce the effect of lactic acid on enamel.

(iv) Over the next 10 days, the composition of plaque varies, with *Actinomyces*, *Veillonella*, *Fusobacteria* and *Bacteroides species* replacing a proportion of the streptococci.

(v) After 10-12 days, spirochaetes and motile Gram-negative rods are present in subgingival plaque.

Structure
In the early stages of formation, plaque has no structural organisation, but with 'maturity' a partially organised structure develops as follows:

(i) 'Cell ghosts' and cocci and bacilli with thick cell walls are common close to the enamel surface.

(ii) Gram-positive filaments run at right angles to the enamel surface and are surrounded by cocci and bacilli.

(iii) Glycogen-like polysaccharide stores are present within some plaque bacteria, e.g. *Strep. mutans*, which can be used when dietary sugar is low — so prolonging the period of acid production

Dental caries
There are three main factors involved in the aetiology of dental caries
1. The tooth and its environment.
2. Diet.
3. Dental plaque.

Theories of caries production

Acidogenic theory: Tooth destruction is caused by decalcification of enamel and dentine by fatty acids (e.g. lactic, formic, propionic) which are produced as a result of the metabolic activity of plaque bacteria using sucrose as substrate.

The proteolytic theory and the proteolytic chelation theory have also been reported but are believed to play a minor role compared with the acidogenic theory.

CARIOGENIC BACTERIA

Cariogenic bacteria are microorganisms which produce dental caries in germ-free animals in the presence of a high carbohydrate (sucrose) diet. The most cariogenic organism is *Strep. mutans*, but *Strep. sanguis*, *Lactobacillus species* and a few other bacteria also give variable positive results.

Streptococcus mutans

Strep. mutans possesses the following properties which make it particularly effective in producing tooth destruction:

a. Fatty acid production (especially lactic) from carbohydrates (especially sucrose).
b. Lowering of the pH to 4.6–4.8.
c. Production of intra- and extracellular polysaccharides.
d. Adherence to enamel and to other oral bacteria.

Further evidence supporting its role in tooth decay is:

(i) It is commonly found in the plaque of carious teeth, but not in the absence of caries.
(ii) It produces caries in experimental animals including primates.
(iii) In animals, vaccines containing *Strep. mutans* reduce the incidence of caries.

Taken together, these findings are strong evidence that *Strep. mutans* is of prime importance in the initiation of caries in experimental animals and man. However, in man, caries can develop in the absence of *Strep. mutans*, so that this cariogenic organism should be regarded as an important but not essential microbial factor in caries.

Lactobacilli

There is a direct correlation between the severity of dental caries and the numbers of lactobacilli present in saliva but this is probably a result rather than an initiating cause of caries. Only small numbers of lactobacilli are present in plaque in early carious lesions but they can be isolated readily in the very low pH conditions of advancing lesions. Lactobacilli cause destruction of dentine.

Diet

A high sucrose diet produces a bulky, adherent plaque which is rich in polysaccharide and has a marked potential for acid production. In a patient fed by stomach tube, a scanty, poorly adherent plaque is formed which is low in polysaccharide and has a low potentional to produce acid. Intermittent intake of small amounts of sucrose is more likely to produce caries than large amounts consumed at one sitting. Dental caries is a disease found mainly in countries with a high sugar consumption. Although virtually unknown in countries with low sugar consumption, the introduction of sugar into the diet causes a dramatic increase in the incidence of caries.

Caries resistance
Relatively few individuals in Western countries are completely free from caries. However, there are five possible reasons for resistance to caries:
    Lack of fermentable carbohydrate in the diet
    Differences in the microbiota of dental plaque
    Protective factors in the saliva
    Teeth which are inherently resistant to caries
    Acquired resistance to caries due to fluoride
Studies on caries-free and caries-active groups have failed to find any common factor or group of factors which could explain the difference in the incidence of caries.

TREATMENT
1. If caries is extensive, tooth extraction, or in some cases, root treatment (i.e. removal of pulp of tooth).
2. If caries is moderate, restoration of tooth with filling material

PREVENTION
Change in dietary habits: e.g. reduction in sucrose intake.
Addition of fluoride to diet, either systemically (1 ppm) via the water supply or topically as a gel. Fluoride ions react with hydroxy groups in the hydroxyapatite of enamel and reduce the solubility of enamel in acid. Fluoride may also act by interfering with bacterial enzymes and so reduce acid production.
Oral hygiene. Toothbrushing alone is not successful in reducing the incidence of caries but in combination with other aids, e.g. dental floss and interdental brushes, some reduction may be achived.
Chlorhexidine: 0.2 per cent aqueous chlorhexidine adheres preferentially to enamel and reduces plaque formation with a consequent protective effect on dental caries.
Fissure sealants: the occlusal surfaces of molar teeth are coated with an adhesive resin which prevents occlusal caries.
Immunisation using *Strep. mutans* has had some limited success in reducing caries in rats and monkeys. However, it is not yet known if these results would also be found in humans; *Strep. mutans* and human

heart muscle may possess shared antigens so that there are some at least theoretical problems in using the vaccine in man.

Chronic gingivitis and periodontal disease

The main factors involved in chronic gingivitis and periodontal disease are subgingival plaque, the gum tissues (gingivae), the tissues which support the tooth in alveolar bone (the periodontium) and the specific and non-specific defence mechanisms of the host.

Chronic gingivitis is an almost universal disease and is due to a persistent low-grade infection with plaque bacteria.

Chronic periodontal disease is characterised by destruction of the periodontal membrane, resorption of bone, migration of the epithelial attachment towards the apex of the tooth, and formation of pockets around the tooth. If untreated, the tooth will eventually loosen and fall out. The incidence and severity of periodontal disease increases with age and is rarely found in children.

The factors involved in the change from chronic gingivitis to destructive periodontal disease — which may take years — are unclear.

Subgingival plaque

The degree of non-specific gingivitis — and to some extent chronic periodontitis — is related to the amount of plaque in the subgingival area. The protein-rich gingival exudate is the main nutrient for bacteria in this region which may explain, at least partially, why *Bacteroides species,* campylobacter and spirochaetes are almost solely found in that site.

Microorganisms

If human strains of oral bacteria, e.g. *Actinomyces viscosus,* are inoculated into the mouth of germ-free animals, gingivitis and peridontitis develop. Antibiotics in the animal diet reduce the incidence of gingivitis and bone loss.

There is a possible relationship between *progressive adult periodontal* disease and *Bacteroides asaccharolyticus* and *Eikenella corrodens.* In addition, spirochaetes and motile Gram-negative bacilli predominate microscopically in areas of periodontal destruction.

Actinobacillus actinomycetemcomitans is an anaerobic Gram-negative bacillus which possesses a toxin which lyses polymorphs and may be aetiologically involved in *juvenile periodontosis* — a rare disease characterised by severe resorption of bone around first molar and incisor teeth.

At present, periodontal disease appears to be the result of a mixed infection with certain bacteria playing dominant roles.

Factors toxic to periodontium

The mechanisms causing destruction of the periodontium are not firmly established but the following factors have been postulated:

1. *Microbial factors:*

   a. *Metabolites* produced by plaque bacteria, e.g. hydrogen sulphide, indole, ammonia and toxic amines.

   b. *Enzymes* e.g. collagenase, hyaluronidase, β-glucuronidase.

   c. *Extracellular polymers,* e.g. high molecular weight dextrans which are resistant to phagocytosis and which in host tissues cause release of lytic enzymes from phagocytic cells.

   d. *Endotoxin:* lipopolysaccharides from Gram-negative bacteria which have a wide range of activity, e.g.

   (i) Cytotoxic (alone or with complement) to fibroblasts.

   (ii) Induce inflammatory response.

   (iii) Chemotactic.

   (iv) Induce bone resorption.

   e. *Fatty acids,* e.g. formic, acetic, propionic, lactic.

2. *Dental calculus.* A mineralised deposit usually derived by the calcification of dental plaque and deposited on teeth either above or below the gum margin. In the absence of viable bacteria, this material does not produce gingivitis or periodontitis in animals.

3. *Phagocytic cells.* Microphages and macrophages are present in the lesions of chronic gingivitis and periodontitis. These cells possess lysosomes which contain a range of lytic enzymes. Various factors, e.g. endotoxin, antigen–antibody complexes and polysaccharides, enhance lysosomal hydrolase synthesis and release which may set up a self-perpetuating and progressively destructive inflammatory reaction in surrounding tissues.

4. *Immune factors.* It is postulated that humoral and cell-mediated immune reactions to plaque microorganism lead to tissue destruction in periodontal disease, but there is no real clinical evidence for this. Little is known about protection by immune mechanisms. All four types of hypersensitivity reaction may play a part in periodontal disease — i.e. type I (anaphylaxis), type II (cytotoxic), type III (immune complex), type IV (delayed hypersensitivity) — and contribute to damage of host tissues.

Factors mediating alveolar bone loss

1. *Complexing agents and hydrolytic enzymes* from plaque may act directly on bone causing decalcification and hydrolysis of its organic matrix.

2. *Antigen–antibody* complexes may activate complement with the synthesis of prostaglandins which stimulate bone resorption.

3. *Endotoxin.*

4. *Cell-mediated immune reactions* produce lymphokines which include an osteoclast activating factor.
5. *Age-related osteoporosis* may also be involved.

Treatment

Chronic gingivitis: scaling and instruction in oral hygiene with surgery in severe cases.

Chronic periodontitis: Scaling, instruction in oral hygiene with surgery to remove non-functional tissue.

Periodontal disease: is usually a slowly progressive disease which if untreated over a period of 15–20 years may require tooth extraction. In view of the many toxic factors which have been described *in vitro* it is surprising that the destruction proceeds so slowly, and it is likely that there are powerful protective factors which delay or inhibit the cytotoxic process.

Periapical abscess (acute local periodontitis)

Clinical features

Spread of infection: via the root canal following death of the dental pulp by caries, produces inflammation of the periodontal membrane around the tooth apex; an abscess forms, which may remain localised or spread into the surrounding bone (osteomyelitis) or into soft tissues (buccal or palatal abscess).

Clinical presentation: varies depending on the virulence and spread of the causative microorganisms and the status of host tissues. Signs and symptoms include pain, swelling and tenderness to tooth percussion.

Cellulitis develops if infection spreads via fascial planes: in the *maxilla,* the cheek and eye may be affected; if the *sublingual or submaxillary area* is involved, swelling of the floor of the mouth and parapharyngeal space may lead to difficulty in breathing.

Microbiology and diagnosis

Causal organisms: are from dental caries lesions, especially α- and non-haemolytic streptococci, anaerobic cocci, *Haemophilus, Veillonella, Bacteroides, Fusobacterium* and *Lactobacillus species;* *Staph. aureus* and β-*haemolytic streptococci* are rarely isolated.

Pus is collected aseptically, and transported to the laboratory for culture, if possible in transport medium with reducing properties.

Treatment

Localised abscess: drainage of pus by extraction of carious tooth, together with antibiotics in severe cases (penicillin, erythromycin or cephaloridine). Alternatively root canal therapy which 'saves' the tooth can be used. If infection spreads, intra- or extra-oral incision of abscess together with tooth extraction and antibiotics may be required.

Acute osteomyelitis

Clinical features
Usually due to local infection from a tooth and its associated structures, e.g. spread from an acute periapical abscess, or from a periodontal pocket when the jaw is fractured.

Predisposing factors are: trauma, leukaemia, severe anaemia, uncontrolled diabetes and immunosuppressive therapy. There may be bone resorption and sequestration. Symptoms include throbbing, deep-seated pain, swelling, trismus, enlarged lymph nodes and pus may exude from a socket or around the necks of teeth.

Microbiology and diagnosis
Pus or a swab from the depths of the lesion is collected. The causal organisms isolated are usually members of the commensal oral flora, e.g. *Bacteroides species,* anaerobic and facultative streptococci and fusobacteria. Rarely, *Staph. aureus* and *Strep. pyogenes* are isolated.

Treatment
1. *Initially, penicillin,* which may be changed subject to bacteriological findings.
2. *Surgical intervention:* drainage and removal of sequestra.

Sialadenitis

Clinical features
Sialadenitis may be acute or chronic.

1. *Acute.* This infection is generally associated with diseases which markedly reduce salivary secretion, e.g. post-operative dehydration, salivary gland destruction due to Sjögren's syndrome or radiotherapy for salivary gland tumour. The clinical presentation includes bilateral or unilateral pain and swelling, malaise and discharge of pus via the duct. Mumps is the most common cause of acute sialadenitis and produces a 'cloudy' non-purulent salivary discharge.

2. *Chronic.* This is more common in the submandibular gland, and usually presents with intermittent pain and swelling of the affected gland. The duct is often obstructed by calculus.

Microbiology and diagnosis

A specimen of pus is collected aseptically, if necessary, by cannulating the duct with a fine plastic catheter and irrigating with saline. The bacteria commonly isolated are:

Acute infection: Staph. aureus and β-haemolytic streptococci.
Chronic infection: α-haemolytic streptococci, *H. influenzae*

Treatment

Acute: Antibiotic therapy with dehydration and surgical drainage if necessary.

Chronic: Antibiotic therapy with surgical removal of calculus if present.

2. SPECIFIC LOCALISED INFECTIONS

Actinomycosis

Clinical features

An uncommon, endogenous, chronic suppurative infection usually involving the soft tissues in the submandibular region and angle of mandible. Bone is rarely affected. Although related to previous trauma — especially tooth extraction — the predisposing factors are unknown. The chronic form presents as a hard indurated swelling with or without discharging sinuses in the neck: it may be preceded by an acute form which is similar clinically to a periapical abscess.

Microbiology and diagnosis

Causal organism: Actinomyces israelii (occasionally *Actino naeslundii*) present in pus in the form of macroscopic aggregates, the so-called 'sulphur granules'. The organism is slow growing and cannot normally be isolated if the provisional diagnosis is not marked on the request form. The source of *Actino israelii* is dental plaque and tonsillar crypts. A fluid sample of pus should be collected if possible or dressings from sinus discharge may be examined for 'sulphur granules'.

Treatment

1. *Acute form.* Drain abscess by external incision, and give short-term antibiotic therapy, e.g. penicillin or tetracycline for about 7 days.
2. *Chronic form.* Establish drainage by surgical intervention and give long-term antibiotic therapy for about 6 weeks.

Candidosis

Clinical features

Oral candidosis is an opportunistic infection which is associated with alteration in the local and systemic defence mechanisms of the host, e.g. physiological, antibiotics, steroids, immunosuppressive drugs, diabetes and leukaemia.

Oral candidosis has a number of *clinical variations*: one group is common, transient and confined to the mouth and easily treated; the converse is true of the other main group which includes patients with rare immune deficiencies.

The common clinical types are:

1. *Acute pseudomembranous candidosis (thrush).* The lesions are soft creamy patches on the oral mucosa which are easily rubbed off leaving a red surface. The disease is found mainly in 'sickly' infants, the elderly and adults with debilitating diseases.
2. *Acute atrophic candidosis.* The oral mucosa especially the tongue is painful, red and swollen. There is usually a recent past history of broad-spectrum antibiotic therapy. Since small areas of 'plaque' can be found, this is probably a variant of 'thrush'.
3. *Chronic atrophic candidosis (denture stomatitis).* The area of mucosa covered by the upper denture is erythematous and is sharply demarcated from the pallor of the normal mucosa. Angular cheilitis may accompany this form of candidosis.

Microbiology and diagnosis

Although *Candida albicans* is commonly isolated from lesions, other species are also involved, e.g. *C. tropicalis* and *C. krusei*: 50 per cent of the population are oral carriers of *C. albicans*. Gram-stained smears from the lesions of 'thrush' always contain pus cells and yeasts in the 'pseudohyphal phase'. Smears are rarely positive in acute atrophic or chronic atrophic candidosis (although in the latter the denture is usually positive). Swabs from lesions are cultured on Sabouraud's medium.

Treatment

The local and systemic predisposing factors should be corrected, and nystatin or Amphotericin B in lozenges and creams prescribed.

Angular cheilitis

Clinical features

Angular cheilitis may be mild with diffuse reddening or severe with ulceration and crusted fissures at the angles of the mouth.

Predisposing factors are: inadequate dentures, iron or folate deficiency, and microbial infection; the disease is often associated with oral candidosis.

Microbiology and diagnosis

The infective component is opportunistic. The microorganisms isolated are usually *C. albicans* or *Staph. aureus* either alone or in combination; *Strep. agalactiae* is occasionally involved. The mouth is the reservoir of candida and the nose of *Staph. aureus*. Smears and moistened swabs are collected from the lesions, and cultured on blood agar and Sabouraud's agar.

Treatment

1. Correct predisposing factors.
2. *Candida* — nystatin or amphotericin B cream and lozenges to treat the intra-oral reservoir and the lesions at the angles of the mouth.
3. *Staph. aureus* — fusidic acid or tetracycline cream to treat lesions and nasal reservoir.
4. Combined antifungal and antistaphylococcal therapy used in mixed infections, e.g. miconazole cream.

Acute ulcerative gingivitis (Vincent's gingivitis)

Clinical features

The disease is mainly found in young adults and is characterised by crater-shaped or punched-out ulcers covered by a greyish slough which develops initially at the tips of the interdental papillae: there is usually a particularly unpleasant halitosis; cervical lymph nodes are sometimes enlarged. The mouth is sore and the gums bleed easily. If untreated the infection *spreads locally* with destruction of the gum tissues and periodontium and becomes chronic with acute exacerbations.

Predisposing factors include: poor oral hygiene, debilitating diseases, heavy smoking, stress. These factors are believed to produce a peripheral vasoconstriction with reduced gingival circulation.

Microbiology and diagnosis

Acute ulcerative gingivitis is a non-transmissible endogenous opportunistic infection. The bacteria associated with the disease are Gram-negative bacilli — *Fusobacterium species* — and a spirochaete, *Borrelia vincentii* (fusospirochaetal complex). Bacteroides and campylobacter are present in relatively small numbers but may play a part in tissue destruction. Smears from the depths of the lesion show numerous leucocytes, spirochaetes and fusobacteria. The differential diagnosis includes primary herpetic stomatitis and, more rarely, acute leukaemia. In some cases acute ulcerative gingivitis is superimposed on both these diseases.

Treatment
Should be started immediately to reduce tissue destruction.
1. Oral hygiene — scaling and irrigation.
2. Metronidazole or penicillin.

Cancrum oris (noma)

In Africa and Asia children may develop acute ulcerative gingivitis which may progress to massive necrosis of the face with perforation of the lips and cheeks with exposure of bone. Predisposing factors are malnutrition, and systemic infections such as measles, malaria and herpetic stomatitis.

Treatment: penicillin, metronidazole.

Vincent's angina

Characterised by sloughing lesions of the tonsils and pharynx. It may or may not be an extension of acute ulcerative gingivitis.

Treatment: metronidazole.

Staphylococcal lymphadenitis

Clinical features
A tender, red submandibular swelling is found in children 1–10 years of age: pain is not marked, and no dental focus of infection is present.

Microbiology
Staph. aureus is readily isolated from pus collected by external incision. Bacteriophage typing confirms that the infection is endogenous — from the anterior nares.

Differential diagnosis: includes tuberculous lymphadenitis which usually produces a slow-growing, indolent, painful swelling.

Treatment
Conservative treatment if infection is at an early stage, using flucloxacillin or erythromycin (most strains are β-lactamase producers). More severe infections require surgical drainage and antibiotics.

3. SPECIFIC SYSTEMIC INFECTIONS WITH ORAL SYMPTOMS

Tuberculosis

Clinical features

Oral tuberculosis lesions: are uncommon and are usually secondary to primary infections in the lung. Many oral sites may be involved but the tongue is the most common: the lesions are usually ulcerated and painful.

Tuberculous lymphadenitis: commonly affects the cervical group of lymph nodes: initially the swelling is firm and mobile but later becomes fixed with abscess and sinus formation. Pain is usually present and the lesions may be unilateral or bilateral, single or multiple.

Microbiology and diagnosis

Causal agent: in almost all oral lesions is *Myco. tuberculosis*. In tuberculosis lymphadenitis 10 per cent of cases are caused by *Myco. bovis* and 15 per cent (mainly in children) by atypical (anonymous) mycobacteria.

Oral lesions:

Ziehl-Neelsen films are usually negative.

Unless the diagnosis is considered, and often it is not, pus or necrotic debris fails to be cultured on Lowenstein-Jensen medium and the opportunity to isolate *Myco. tuberculosis* is missed. As a result the true nature of the oral lesions is diagnosed by histopathology of a biopsy specimen.

Lympadenitis. If caseation has taken place pus is available for examination.

Ziehl-Neelsen films: often demonstrates acid and alcohol fast bacilli.

Lowenstein-Jensen culture: positive for *Myco. tuberculosis* in 4–6 weeks.

Treatment

As for pulmonary tuberculosis. Atypical mycobacteria may be resistant to routine antituberculous therapy.

Syphilis

Clinical features

The oral manifestations of syphilis — now an uncommon disease — are:

1. *Primary chancre.* The lip, angle of mouth or tongue are rare sites for a primary lesion: these extragenital chancres are infectious. The cervical lymph nodes are enlarged.

2. *Secondary* (mucous patches). The lesions are slightly raised — greyish-white, glistening patches on the mucous membrane of the tonsil, tongue and cheek: they may coalesce to form 'snail track' ulcers, which are infectious.

3. *Tertiary (gumma)*. Common sites in the mouth are the hard and soft palate and tongue. A gumma starts as a small pale raised area, which ulcerates and progresses to a large area of necrosis with bone destruction. Lesions have low infectivity.

4. *Congenital*. Dental lesions of congenital syphilis are the result of infection of developing tooth germs by *Tr. pallidum*. The permanent dentition is usually affected, appearing clinically as deformed teeth, e.g. 'mulberry molars', Hutchinson's incisors.

Microbiology and diagnosis
Dark-ground microscopy of oral lesions to demonstrate *Tr. pallidum* is usually inconclusive due to the presence of commensal spirochaetes. Best diagnosed by serology.

Treatment
See Chapter 40.

Gonorrhoea

Clinical features
Oropharyngeal gonorrhoea is being increasingly recognised in sexually active adults. The lesions are usually a primary infection but sometimes due to autogenous spread from genital lesions.

The clinical presentation varies from mild, asymptomatic pharyngitis to an acute disease with vesicles, ulcers and pseudomembranes affecting gums, buccal mucosa and pharynx.

Microbiology and diagnosis
Casual agent: *N. gonorrhoeae*. Gram-stained smears of oral lesions are of little value due to the presence of commensal Neisseria. Swabs from the lesions are sent in transport medium but isolation and identification may be difficult due to the presence of commensal neisseria.

Treatment
Penicillin or tetracycline. Spectinomycin is only effective in curing 50 per cent of patients with oropharyngeal lesions.

4. SYSTEMIC INFECTIONS WITHOUT ORAL SYMPTOMS BUT OFTEN CAUSED BY ORAL COMMENSAL BACTERIA

Infective endocarditis is a serious illness resulting in destruction of heart valves which are, as a rule, already abnormal due to pre-existing disease. It has been estimated that about one in eight cases of infective endocarditis arise from the transient bacteraemia which follows dental procedures such as tooth extraction, scaling, periodontal surgery and endodontics (root canal fillings).

43

Zoonoses

Zoonoses are infections acquired from animals. Not surprisingly, many of those affected are agricultural workers and veterinary surgeons, but the general public are also at risk for example, through contaminated meat and milk.

Table 43.1 lists the main bacterial zoonoses. Other zoonoses are caused by rickettsiae and viruses.

Table 43.1 Bacterial zoonoses

Disease	Causal organism	Main animal host
Food-poisoning	*Salmonella species*	Cattle, poultry
Food-poisoning	*Campylobacter species*	Poultry, other domestic animals
Tuberculosis	*Mycobacterium bovis*	Cattle
Brucellosis	*Brucella abortus*	Cattle
	Brucella melitensis	Goats, sheep
	Brucella suis	Pigs
Anthrax	*Bacillus anthracis*	Cattle
Plague	*Yersina pestis*	Rats
Mesenteric adenitis, enteritis	{ *Yersina pseudotuberculosis* *Yersina enterocolitica*	Various animals
Septic dog bite	*Pasteurella multocida*	Dogs and cats
Tularaemia	*Francisella tularensis*	Squirrels, rodents
Leptospirosis	*Leptospira interrogans*	Rats, pigs, dogs, cattle
Listeriosis	*Listeria monocytogenes*	Various domestic and wild animals
Erysipeloid	*Erysipelothrix rhusiopathiae*	Pigs, other animals and fish
Rat bite fever	{ *Streptobacillus moniliformis* *Spirillum minus*	Rats, mice

Domestic animals are more likely to be sources of infection than wild animals due to their closer contact with people. Table 43.2 shows the principal sources and routes of infection acquired from domestic animals.

Table 43.2   Domestic animals: sources and routes of infection

Source	Route	At risk
Infected animals	Contact	Farm workers, veterinary surgeons, slaughtermen
Contaminated pastures, straw, dust, soil	Inhalation, contact	Farm workers, veterinary surgeons
Milk	Ingestion	People drinking unpasteurised milk
Meat	Ingestion	Anyone eating meat
Hides, bones, other animal products	Contact	Industrial workers handling animal products, occasionally general public
	Inhalation	Industrial workers handling animal products

Salmonella and campylobacter food-poisoning are by far the most common zoonoses in Britain: these diseases and tuberculosis due to *Mycobacterium bovis* are described in Chapters 30 and 38 respectively.

BRUCELLOSIS

Although brucellosis in Britain is being brought under control it remains worldwide in distribution: the Causal *Brucella species* are named after Bruce who discovered the cause of one form of the disease while serving as an army doctor in Malta.

Causal organisms — brucellae — are small Gram-negative coccobacilli: the main species are listed in Table 43.3.

Table 43.3   Brucella species

Species	Animal host	Geographical distribution
Brucella abortus	Cattle	Worldwide
Brucella melitensis	Goats and sheep	Mediterranean area
Brucella suis	Pigs	USA, Denmark

CLINICAL FEATURES

Incubation period: 1 week to 6 months.

Signs and symptoms: undulant fever, a prolonged debilitating febrile illness with remissions and relapses, often becoming chronic when it persists for months or even years: sweating, anorexia, constipation, rigors, weakness and lassitude are the main symptoms; the spleen is often enlarged, there may be arthritis, orchitis and neuralgia; acute brucellosis is a *septicaemic* illness: abortion is not a feature of human brucellosis.

In chronic brucellosis the symptons are of vague ill-health: 'psychiatric morbidity' (usually depression) may overshadow physical symptoms.

Duration: on average 3 months: even without therapy symptoms of acute brucellosis usually disappear within one year.

Severity: brucellosis due to *Br. melitensis* tends to be a more severe disease than that due to *Br. abortus.*

Mortality: even before antibiotics, brucellosis had a low case fatality rate — around 2 per cent.

Brucella suis infection is less common than that due to *Br. abortus* or *Br. melitensis.*

PATHOGENESIS

Route of infection: usually by drinking unpasteurised *contaminated milk:* farm workers and veterinary surgeons are not uncommonly infected by *direct contact* with infected animals or their products — especially the placenta or uterine discharges from parturient animals.

Spread in the body is via lymph channels to lymph nodes and blood stream: the organism then becomes widely distributed in organs and tissues to produce the symptoms of acute brucellosis.

Intracellular parasitism: brucellae have a particular tendency to persist intracellularly notably in spleen, liver and lymph nodes; this is the reason for the notorious difficulty of eradicating the infection by antibiotic therapy. Antigen release from these sites may lead to an immune complex syndrome and cause the symptoms of chronic brucellosis.

DIAGNOSIS

Isolation

Specimens: blood culture in liver infusion or glucose serum broth incubated in CO_2.

Observe for growth: retain cultures for six weeks.

Identify: by dye test, H_2S production, urease activity and agglutination with monospecific antisera (*Br. abortus* and *Br. melitensis* share antigens).

Positive blood cultures make certain the diagnosis of acute brucellosis but numerous sets should be taken because the organism is notoriously difficult to isolate. *Br. melitensis* is more readily cultured than *Br. abortus.*

Serology:

Detect antibody levels against both *Br. abortus* and *Br. melitensis*. Confusingly, some biotypes of *Br. abortus* have a preponderance of melitensis antigen with a minor content of abortus antigen: infection with *Br. abortus* may therefore result in higher titres of antibody to *Br. melitensis* than to *Br. abortus*.

1. *Direct (standard) agglutination test.* Dilutions of the patient's serum are tested against a brucella suspension and examined for agglutination: *prozones* (i.e. absence of agglutination in low dilutions of serum which contain high levels of antibody) are common.

2. *Indirect agglutination test.* Incomplete antibodies which combine with the organisms in the brucella suspension but are unable to agglutinate them may be found in brucellosis: detect by *Coombs' test.*

   Coombs' test: tubes from a negative direct agglutination test are centrifuged to deposit the bacteria; antihuman globulin is added and if incomplete antibody is present which has coated the bacteria agglutination will now take place.

3. *Complement fixation test:* useful in detecting incomplete antibody: widely used because easier to perform than Coombs' test.

4. *Radioimmunoassay:* not generally available but the most sensitive and best test; the only test that can distinguish IgM, IgG and IgA brucella-specific antibodies with certainty.

Immunoglobulins involved in the main serological tests for brucellosis are shown below together with the titres regarded as significant.

Table 43.4 Tests for brucella antibodies

	Direct agglutination test	Indirect agglutination test	Complement fixation test
Immunoglobulin class responsible for a positive test	Mainly IgM but also IgG and IgA	Mainly IgG but also IgA and IgM	Mainly IgG but also IgM *not* IgA
Significant titre	⩾80	⩾80	⩾16

Acute brucellosis: at the time of presentation antibody levels are almost always high — direct agglutination titres of 1000 or more are common: IgM and IgG are raised and titres decline slowly to low levels or zero with clinical recovery.

Chronic brucellosis: levels of IgG and IgA but not IgM are elevated: the indirect agglutination test and the complement-fixation test may be positive and the direct agglutination test negative.

Similar findings are encountered in healthy individuals whose immunity is being repeatedly stimulated by contact with brucella organisms and 'positive' serological tests are common in those at occupational risk.

Note. It is impossible to distinguish by serological tests chronic active brucellosis from seropositive but inactive infection — especially in rural areas where exposure to infection is common.

The administration of tetracycline (a 'therapeutic test') may result in clinical improvement and a fall in antibody levels.

Delayed-type hypersensitivity: the *brucellin* skin test is similar to the tuberculin test and indicates present or past infection with brucellae: often positive in healthy people from agricultural areas. Of little diagnostic value — in fact, the test may stimulate antibody production and make assessment of serological tests even more difficult.

TREATMENT

Tetracyclines: but prolonged treatment is necessary, e.g. for 4–6 weeks: streptomycin may be given in addition for three weeks of the course.

Cotrimoxazole: promising results in preliminary trials.

BRUCELLOSIS IN ANIMALS

Animal brucellosis is also a chronic debilitating, septicaemic disease: animals are particularly infectious at parturition because of the heavy contamination of placenta and products of conception. *Erythritol,* which is contained in placental tissue, is a growth factor for brucellae.

Abortion is a common sequel of brucellosis in cattle and can result in serious economic losses in herds.

Milk: brucellae are usually excreted in the milk of infected animals.

CONTROL

Brucellosis is well on the way to becoming eradicated in cattle in Britain.

Vaccination: of calves between the 4th and 6th month with a live attenuated vaccine (S19) results in significant protection from infection and abortion. In Britain vaccination has been abandoned in favour of eradication.

Eradication:

1. *Infected herds* are recognised by detecting antibodies to *Br. abortus* in pooled milk samples. The milk ring test is used.

Method: add a concentrated suspension of *Br. abortus* stained with haematoxylin to the milk and centrifuge.

Observe: in a positive test the development of a deep blue ring in the cream layer.

Mechanism: antibodies agglutinate the stained bacteria which rise to the surface with the fat globules.

2. *Individual infected animals* are then identified by isolating *Br. abortus* directly from cream or demonstrating antibodies in their serum or milk whey: they must then be slaughtered.

Accredited brucellosis-free herds are now being built up by this policy.

ANTHRAX

Anthrax is a disease of animals from which man is only occasionally infected: although a wide variety of animals are susceptible, anthrax is mainly a disease of herbivores and especially cattle.

Causal organism: Bacillus anthracis —a large sporing, capsulated, Gram-positive bacillus.

CLINICAL FEATURES

Cutaneous anthrax

Cutaneous anthrax, or malignant pustule, is due to direct inoculation of the skin from infected animals or animal products: an inflamed but painless lesion with surrounding oedema and a characteristic black eschar: local lymph nodes are usually enlarged: if untreated, it may progress to a septicaemia with death from overwhelming infection.

Acquired in two ways:

1. *Industrial:* in leather workers, tanners, workers in bone meal factories; malignant pustule of the neck and shoulders was an occupational hazard of hide porters due to rubbing of infected hides carried on their backs.

2. *Non-industrial:* seen in people who work with animals (an occupational hazard in agriculture) and occasionally in the general public due to handling of infected shaving brushes, leather goods and clothes, bone meal, etc.

Pulmonary anthrax

Also called *woolsorter's disease* — a name indicating its mode of spread by inhalation of spores in workers handling contaminated wool: now a rarity — a disease of the nineteenth century.

Clinically: a severe disease with a high mortality rate: in fact virtually always fatal before antibiotics: characterised by fever, increasing respiratory distress and death.

Gastrointestinal anthrax

Also a lethal disease; due to ingestion of *B. anthracis* or its spores; fortunately very rare.

PATHOGENESIS

Long regarded as a classic example of a disease purely due to the invasive properties of the causal organism, *B. anthracis* is now known to produce its effects also by the formation of three toxins:

1. Oedema factor.
2. Protective factor (antibody to this is responsible for immunity to the disease).
3. Lethal factor.

DIAGNOSIS

Specimens: swab or sample of exudate from malignant pustule: sputum from suspected pulmonary anthrax.

Examination:

1. *Direct film:* Gram stained.
   Observe for typical large Gram-positive bacilli.
2. *Culture:* on ordinary media.
   Observe: for growth of typical 'curled hair lock' colonies.
   Confirm identity of isolated bacilli by subcutaneous inoculation of guinea pigs.
   Observe: local lesion with typical gelatinous oedema and death from septicaemic infection.

TREATMENT

Penicillin.

ANIMAL ANTHRAX

A rapidly fatal, septicaemic infection in which huge numbers of organisms are present in the blood and are shed in discharges from the body orifices.

Pastures used by infected animals become contaminated with anthrax spores and may remain infective for many years.

DIAGNOSIS

Specimen: blood from a sick animal: an ear cut off post-mortem and sent to the laboratory (packed with due precautions).

Examine by:

1. *Direct demonstration:* of large Gram-positive bacilli — usually very numerous — in a blood smear: *B. anthracis* does not form spores in the animal body. *McFadyen's reaction:* the demonstration of pink-staining capsular deposits lying between the bacilli in a blood film stained by methylene blue.
2. *Culture* on ordinary media for typical colonies.

CONTROL

Animals suspected of being infected must be notified to the authorities. Dead animals must be disposed of by burying in quick-lime; a careful check must be kept on the rest of the herd.

Imported animal products are also subject to inspection and control.

PLAGUE

Plague is a natural disease of rats which occasionally spreads to man via the bite of infected fleas: one of the great epidemic diseases, it was the Black Death (which killed millions in the fourteenth century) and the Great Plague of London and elsewhere in Britain during the mid-seventeenth century. Plague still exists in endemic foci in the Western USA, South America, Africa and the Far East.

Causal organism: Yersinia pestis, a small Gram-negative bacillus.

CLINICAL FEATURES

There are two forms of plague.

1. Bubonic plague

The commoner form; a septicaemic illness with fever, prostration, mental confusion and characteristic enlargement with profuse pus formation of inguinal glands (buboes): high mortality rate — about 50 per cent in untreated cases; not contagious — acquired by the bite of fleas.

2. Pneumonic plague

A rare but highly infectious form of plague affecting the lungs; the route of infection is by inhalation of infected respiratory secretions: virtually always fatal.

Reservoir is mainly rural or sylvatic rats: most human epidemics have been due to spread of the disease to urban rats.

Vector: the rat flea *Xenopsylla cheopis:* plague spreads amongst rats via infected fleas; rat fleas do not usually infest man but do so if the rat population has been killed off by the disease.

Infected fleas transmit the disease during biting: the bacilli multiply in the flea including the proventriculus from where the organisms are injected when the flea bites and sucks blood again.

EPIDEMIOLOGY

The disease may become epidemic in the rat population causing heavy mortality: when there are insufficient rats, the fleas turn to man and spread the disease through human populations.

DIAGNOSIS

Isolation

Specimen: bubo pus.

Gram film: examine for Gram-negative bacilli darker stained at the ends than in the middle (bipolar staining).

Culture: ordinary media.

Identify: biochemical and serological reactions.

TREATMENT

Tetracyclines.

CONTROL

The prevention of rats coming ashore from ships: although there seems little danger of epidemic plague nowadays, precautions and sampling regulations remain in force.

INFECTIONS WITH OTHER YERSINIA SPECIES

Yersinia enterocolitica ⎱ cause enteric infection with
Yersinia pseudotuberculosis ⎰ mesenteric adenitis (which may mimic appendicitis), diarrhoea and fever.

Both organisms are widespread in various animal species: *Y. pseudotuberculosis* causes the disease pseudotuberculosis in animals, but animal infection with *Y. enterocolitica* is usually symptomless.

PASTEURELLA

Pasteurella multocida also known as *P. septica:* a not uncommon cause of sepsis after dog or cat bite.

TULARAEMIA

A plague-like disease spread from ground rodents such as squirrels, in U.S.A., U.S.S.R. and elsewhere: due to *Francisella tularensis.*

LEPTOSPIROSIS

A disease of rodents and sometimes domestic animals transmitted to man by direct or indirect contact with animal urine. Due to occupational risk human disease predominantly affects males.

Causal organism: Leptospira interrogans. This species is divided into 16 serogroups, three of which, canicola, hebdomadis and icterohaemorrhagiae, are responsible for almost all human cases in Britain.

CLINICAL FEATURES

An initial septicaemic phase lasting 3–7 days which presents as an influenzal illness; the diagnosis is rarely made at this stage. Followed by an immune phase when leptospires have disappeared from the blood and antibodies begin to appear, characterised by signs of meningeal irritation (headache, vomiting), conjunctival suffusion and evidence of renal involvement (proteinuria). Leptospirosis is a recognized cause of *aseptic (lymphocytic) meningitis.*

Severe leptospirosis (*Weil's disease*), associated with jaundice, haemorrhages and serious renal damage is uncommon and usually due to icterohaemorrhagiae infections. It is likely that many mild cases of leptospirosis, probably canicola infections, are never diagnosed.

SOURCES OF INFECTION

Canicola infections
Natural host: pigs and dogs.
At risk: those who tend pigs and dogs (especially sick pups). The infection is widespread in dogs.

Hebdomadis infections
Natural host: field mouse; cattle are secondarily infected from pasture or fodder contaminated with mouse urine.
At risk: farmers, often infected from cow urine during milking: a subtype of *L. hebdomadis* — *sejroe* — was the commonest recognised cause of leptospirosis in Britain in 1978.

Icterohaemorrhagiae infections
Natural host: brown rat.
At risk: miners, sewage workers, agricultural workers, fish handlers and others subject to occupational exposure by employment in a moist environment contaminated with rat urine.

Entry of the pathogen is through skin cuts and abrasions; sometimes via the nasopharynx following bathing or accidental immersion in infected water, usually stagnant ponds or canals.

DIAGNOSIS

Isolation
Difficult and rarely accomplished.
  Specimens:
Blood during the first week of illness
Urine (the sample *must* be fresh) during the second and third weeks
Inoculate and observe:

  (i) a suitable serum-enriched liquid medium — growth of leptospires
  (ii) guinea-pigs intraperitoneally — jaundice and death

Serology
The usual method of diagnosis: antibodies are not present in the first week of illness but as a rule can be detected in the second and third weeks.
1. *Agglutination-lysis (Schuffner) test:* using a live suspension of leptospires: serogroup-specific; therefore, dilutions of serum have to be tested against laboratory-maintained cultures of at least three main British serogroups of leptospires. A titre of 100 or greater is regarded as significant.
2. *Complement fixation test:* with a genus-specific antigen (e.g. the saprophytic patoc strain of *L. biflexa*) can detect antibodies to any of the serogroups.

TREATMENT

Penicillin; antibiotics are apparently of limited value in leptospirosis and only modify the disease if given early.

LISTERIOSIS

A rare disease although the causal organism is widely distributed in nature — in soil, silage, water and various animal species.

Causal organism: Listeria monocytogenes, a diphtheroid-like Gram-positive bacillus.

CLINICAL FEATURES

Neonates are the main age group affected, infection being acquired as a result of mild or inapparent infection in the mother.

The disease presents in two main forms:
1. Meningo-encephalitis.
2. Septicaemia with enlarged lymph glands and widespread granulomatous nodules in the viscera — *granulomatosis infantiseptica.*

Adults over the age of 40 years — and particularly the elderly — can also develop the disease usually as meningo-encephalitis.

Mild or inapparent infection is probably not uncommon in the general population: in a pregnant woman there is a risk of transmission to the child.

DIAGNOSIS

Isolation
Specimens: blood cultures, CSF, swabs from genital tract.

Culture: onto blood agar.

Observe: small colonies surrounded by a narrow zone of β-haemolysis.

Identify: small Gram-positive bacilli like corynebacteria but actively motile when grown in broth at 25°C.

TREATMENT

Ampicillin; chloramphenicol.

ERYSIPELOID

An inflammatory lesion of the skin usually of the fingers and hand resembling erysipelas but due to a Gram-positive bacillus — *Erysipelothrix rhusiopathiae:* the cause of swine erysipelas but found in other animals, birds and fish.

Occupational hazard of meat and fish handlers, veterinary surgeons.
Treatment: penicillin.

RAT BITE FEVER

Includes two separate rare diseases transmitted to man by rat bites:
sometimes seen in laboratory workers handling experimental rats.

Causal organism: either (i) *Streptobacillus moniliformis* or (ii)
Spirillum minus, a spiral organism.

44

Infection in compromised patients

Increased susceptibility to infection in certain clinical conditions has been recognised for a very long time, e.g. tuberculosis and staphylococcal skin sepsis are unduly common in diabetes, and when there is an obstructive element in the pathological process such as a kidney stone or a bonchial carcinoma, infection is the rule.

However, modern medicine has resulted in an increasing number of patients who survive because of advances in the management of their disease: they may present special infection problems to both clinicians and microbiologists.

The microorganisms responsible include recognised pathogens and those considered non-pathogenic in the normal host.

The infections, sometimes called *'opportunistic'* are often unusually severe and may present unfamiliar manifestations.

The patients are unusually susceptible because of some alteration in their normal defence mechanisms against infection: i.e. they are *compromised* either by immunodeficiency or by interference with other body defence mechanisms.

The source of infection is either endogenous or exogenous — the latter often due to organisms acquired from the hospital environment.

IMMUNOCOMPROMISED PATIENTS

Deficiencies of the immune response may affect antibody production, cell-mediated immunity or both. There are three main causes:

1. Disease
Many diseases depress the immune response: three of the best studied examples are

  (i) Neoplasms of the lymphoid system: leukaemia, lymphoma (e.g. Hodgkin's disease), multiple myeloma etc.

  (ii) End-stage renal failure.

  (iii) Splenectomy.

2. Therapy

Therapy which depresses — or occasionally abolishes — immune function is now widely used. This includes:

Immunosuppressive drugs
Steroids
Cytotoxic drugs
Radiotherapy

3. Congenital

Rarely, children are born with congenital deficiency of the immune system: this may involve

a. *Immunoglobulin synthesis:* e.g. B-cell deficiency with depressed production of immunoglobulins. All immunoglobulins may be affected as in the Bruton type of hypogammaglobulinaemia or only some, as in hereditary telangiectasia with deficient IgA and IgE.

b. *Cell-mediated immunity:* T-cell deficiency, e.g. thymic hypoplasia (Di George's syndrome).

c. *Combined immunodeficiency:* lack of differentiation of the common lymphoid stem cell resulting in both B- and T-cell deficiency, e.g. Swiss-type agammaglobulinaemia in which there are no lymphocytes or plasma cells in lymphoid organs and the thymus is very small.

d. *Neutrophil function:* several syndromes are recognised which affect different aspects of phagocytosis, e.g. chronic granulomatous disease, Chediak-Higashi syndrome.

Immunosuppressive therapy

After transplantation of kidneys or other organs, patients are maintained for long periods of time on an immunosuppressive regimen specifically designed to reduce the cell-mediated immune response which causes graft rejection. Infection is a particular problem in renal-transplant patients: about one third of the deaths in these patients are due to infection.

Table 44.1 shows the principal causes and results of infection in immunocompromised patients.

PATIENTS IN SPECIAL CARE UNITS

The management of these patients often creates unnatural portals of entry for microorganisms by breaching the normal non-specific defence mechanisms.

Table 44.1 Causes and results of infection in immunocompromised patients

Type of infectious agent	Main microorganisms involved	Clinical manifestations of infection
Bacteria	Escherichia coli Klebsiella species Pseudomonas aeruginosa and other coliforms	Urinary infections, sepsis of colonic origin (e.g. ischiorectal abscess), pneumonia, septicaemia, meningitis
	Mycobacterium tuberculosis	Pulmonary, miliary tuberculosis
	Staphylococcus aureus	Pneumonia, septicaemia
	Streptococcus pneumoniae	Pneumonia, septicaemia, meningitis
	Listeria monocytogenes	Arthritis, meningitis
	Nocardia asteroides	Pneumonia; metastatic abscesses, e.g. brain
Fungi	Candida albicans	Local thrush, systemic candidosis
	Cryptococcus neoformans	Meningoencephalitis
	Aspergillus; especially A. fumigatus Mucor species	Pulmonary, occasionally disseminated infections
Viruses	Herpes simplex	Severe cold sores
	Varicella-zoster	Zoster, sometimes generalised zoster
	Cytomegalovirus	Pneumonitis
	Vaccinia	Generalised vaccinia, chronic progressive vaccinia
	JC (human polyoma) virus	Progressive multifocal leucoencephalopathy
Protozoa	Toxoplasma gondii	Severe toxoplasmosis with involvement of retina and brain
	Pneumocystis carinii	Interstitial pneumonia

Indwelling catheterisation of the urinary tract, an old-established procedure, is associated with a very high incidence of infection: the longer the catheter remains *in situ* the more certain it is that infection will develop.

Modern intensive care subjects patients on artificial ventilation to prolonged intubation of the respiratory tract: in many cases their tracheal secretions soon contain large numbers of coliforms (e.g. *Escherichia coli*, *Klebsiella species*, *Pseudomonas aeruginosa*, *Acinetobacter species*) that are not usually associated with respiratory infection. The significance of these organisms in the respiratory tract is often difficult to interpret: sometimes their presence is merely due to abnormal colonisation but on other occasions they cause tracheobronchitis or pneumonia.

An intravenous (or intra-arterial) catheter constitutes a foreign body in an open wound at a site (the skin) which is normally colonised with bacteria. Long-line intravenous catheters to monitor the central venous pressure or to allow total parenteral nutrition often remain in position for many days. The catheter soon develops a fibrin sheath which can become infected with a variety of microorganisms derived either from the skin flora or from bacteraemia by trapping bacteria circulating for short periods of time in the blood: infection of catheter sites may be the source of a septicaemia.

PATIENTS WITH PROSTHESES

In recent years it has become standard surgical practice to implant complex metal and plastic prostheses (i.e. foreign bodies) at sites deep within the patient — usually with dramatic beneficial results: unfortunately these devices sometimes fail and this is often because of infection.

Orthopaedic prostheses (e.g. hip and knee joints): the bacteria responsible for infection are often skin commensal flora, i.e. *Staphylococcus epidermidis* and/or corynebacteria.

Artificial heart valves: the agents that cause infection on these valves include *Staph. epidermidis*, *Candida albicans* and other microorganisms not usually associated with infective endocarditis.

45

Infections in general practice

Infections are a large part of the general practitioner's workload. Not surprisingly, the diseases seen tend to be somewhat different from those dealt with in hospitals. Although laboratory services are becoming increasingly available to family doctors, most must diagnose and treat patients without laboratory investigations. Antibiotic therapy, therefore, is usually prescribed on a 'best-guess' basis.

This chapter describes the types of infection seen in general practice and indicates how often these are encountered. The figures quoted are from a general practice in Redcar, Yorkshire, situated in an expanding housing estate. Many of the patients work in large steel or chemical plants. The data are reproduced with permission from 'Towards Earlier Diagnosis in Primary Care' by K. Hodgkin, Fourth Edition, Churchill Livingstone, 1978.

RESPIRATORY INFECTIONS

A quarter of a general practitioner's workload is respiratory infection. These infections are therefore not only an enormous medical problem but significantly affect the economy due to the large number of days lost from work and also to the considerable cost of the drugs prescribed for their treatment. Most are viral.

The relative frequency of some of these infections is shown in Table 45.1.

Table 45.1 Respiratory infections in general practice

Syndrome	Cases per 1000 NHS patients per year
Tonsillitis	76
Common Cold	68
Pharyngitis	45
Acute bronchitis	40
Chronic bronchitis	9
Lobar pneumonia	5
Pulmonary tuberculosis	1

The figures in Table 45.1 illustrate the extraordinary frequency of upper respiratory tract infection in the community. The incidence of common colds is, in fact, an underestimate since many patients with them do not consult their doctor: common colds, acute bronchitis and more than half the cases of pharyngitis are due to viruses.

Influenza

The respiratory disease which makes the biggest impact on the community is influenza: epidemics recur every year or two, sometimes annually. When the 'Asian' influenza A virus appeared in 1957, 18 per cent of the adult population became ill: during this epidemic and on the subsequent reappearances of this virus over the following 12 years, a third of the adult population in the practice were affected.

OTHER INFECTIONS

Table 45.2 shows some of the other most common infections dealt with by family doctors.

Table 45.2 Other infections in general practice

Disease	Cases per 1000 NHS patients per year
Gastroenteritis	78
Acute otitis media	44
Cystitis	15

Like common colds, many cases of gastroenteritis are never reported to the general practitioner so that this figure too is almost certainly an underestimate. It is often easy for hospital doctors — and microbiologists — to forget how common are acute infections of the ear especially in children.

CHILDHOOD FEVERS

Most of the acute fevers of childhood are viral: it is of interest to compare their incidence rates to those of the common infections listed above: children — like the elderly — are seen more often by their doctors than other sections of the population.

Table 45.3 shows the incidence of some fevers — estimated in *epidemic years*.

Table 45.3   Incidence of childhood fevers

Childhood fever	Cases per 1000 NHS patients per (epidemic) year
Rubella	60
Mumps	50
Measles	40
Varicella	33
Hepatitis A	12

RARE INFECTIONS

Table 45.4 lists the incidence of some of the uncommon infections seen in general practice over *a 10-year period.*

Table 45.4   Rare bacterial diseases

Disease	Cases seen per 1000 patients in 10 years
Tuberculosis of bones and joints	5
Renal tuberculosis	2
Tuberculous meningitis	1
Typhoid and paratyphoid fever	2
Brucellosis	0.1
Diphtheria	0

These extremely low figures help to set some of the more severe bacterial infections in proportion in terms of community morbidity: doctors who — like the authors of this book — work only in hospital can easily overestimate their importance and their frequency in the community.

FURTHER READING
Towards Earlier Diagnosis in Primary Care. Keith Hodgkin. Fourth Edition. Churchill Livingstone. Edinburgh, 1978.

Treatment and prevention of bacterial disease

46

Antimicrobial therapy

Bacterial infections are among the few diseases in medicine for which specific therapy is available. Despite this, infections are still common.

ANTIBIOTICS

More than a century ago, Pasteur observed that the growth of one microorganism could be inhibited by the products of another. However, most of these early products or 'antibiotics' were toxic to mammalian as well as bacterial cells and were therefore of no therapeutic use. Penicillin, discovered in 1929 but not available for clinical trials until 1940, is the product of a mould, *Penicillium notatum,* and was the first antibiotic drug: it is still the best. Other antibiotics are the products of soil streptomycetes and bacteria of the genus *Bacillus.* Many recently introduced antibiotics, e.g. the semisynthetic penicillins and cephalosporins, have been prepared by the chemical manipulation of existing drugs.

Selective toxicity is the ability to kill or inhibit the growth of a microorganism without harming the cells of the host: an essential requirement for any successful antibiotic.

Chemotherapeutic agents e.g. sulphonamides, trimethoprim and many of the antituberculous drugs, are synthetic drugs with this selective toxicity.

Antimicrobial therapy is most likely to be successful if the infecting organism is sensitive to the drug chosen. This can be achieved if a sound knowledge of microbiology indicates the most likely pathogen in a given condition, i.e. on a 'best-guess' basis: it is better still if the infection is investigated bacteriologically by culture and in vitro sensitivity testing.

Antimicrobial drugs are often classified as *bactericidal* when they kill the infecting bacteria or *bacteriostatic* when they prevent multiplication but do not kill the bacteria: this classification is not always clear cut and may be dose dependent. Apart from a few conditions notably infective endocarditis, appropriate bacteriostatic drugs give excellent therapeutic results.

This chapter deals with the main antibiotics and antimicrobial agents in current use from the point of view of a clinical bacteriologist. Pharmacokinetics are not considered.

The principal drugs are listed in Table 46.1. Subsequent tables summarise information about particular antibiotics including their clinical use: inevitably they are oversimplified and should be used only as a guide to antibiotic therapy.

Table 46.1 Antibiotics and other antimicrobial agents

Drug	Principal use
Penicillins	Wide variety of infections
Cephalosporins	Mainly second-line drugs
Cotrimoxazole	Urinary and chest infections
Aminoglycosides	Severe infections with coliforms
Tetracyclines	Chest infections
Metronidazole	Anaerobic infections
Macrolides	Second-line drugs to benzylpenicillin
Chloramphenicol	Typhoid fever, meningitis
Fusidic acid	Staphylococcal infection
Nalidixic acid	Urinary infection
Nitrofurantoin	Urinary infection
Polymyxin	Infections with *Pseudomonas aeruginosa*
Isoniazid	Tuberculosis
Rifampicin	Tuberculosis
Ethambutol	Tuberculosis

Penicillins
Chemical structure:

β-lactam ring

Mode of action: bactericidal; inhibits cell wall synthesis by combining with the transpeptidase responsible for cross-linking of the peptidoglycan; activity depends on an intact β-lactam ring.

Resistance: is common and is due to production by bacteria of an enzyme β-*lactamase* which inactivates penicillin by acting on the β-lactam ring: in many cases bacterial β-lactamases are plasmid-coded.

Antibacterial spectrum: varies depending on individual penicillins: the activity is determined by the side chains of the penicillin nucleus.

The principal penicillins and their antibacterial action are shown in Table 46.2.

Table 46.2 The penicillins

Penicillin	Admin- istration	Antibacterial spectrum	Clinical use
Benzylpenicillin (Penicillin G)	i.m., i.v.	} Gram-positive bacteria Neisseria	Streptococcal, pneumococcal infection; sensitive staphylococcal infection; clostridial infection; meningitis, gonorrhoea, syphilis, anthrax, actinomycosis
Phenoxymethyl penicillin (Penicillin V)	oral		
Ampicillin Amoxycillin }	oral i.m., i.v.	Similar to penicillin but in addition, *Streptococcus faecalis,* almost all *Haemophilus influenzae* and many coliforms	Urinary and respiratory infections, enteric fever; in combination with other drugs in severe systemic infections
Methicillin Cloxacillin Flucloxacillin }	i.m., i.v. oral i.m., i.v.	Similar to penicillin but less active drugs. Stable to staphylococcal β lactamase	Staphylococcal infections
Carbencillin Ticarcillin }	i.m., i.v.	Similar to ampicillin but in addition, *Pseudomonas aeruginosa* and most *Proteus species* Ticarcillin has activity against Bacteroides	Urinary, respiratory, burns and other infections due to sensitive bacteria especially *Ps. aeruginosa;* severe sepsis usually in combination with other drugs
Azlocillin	i.v.	Coliforms, especially *Ps. aeruginosa*	As for carbenicillin and ticarcillin
Mecillinam	oral i.m., i.v.	Coliforms; low activity against Gram-positive bacteria	Urinary infections; Salmonellosis including enteric fever but further evaluation required

i.m. = intramuscular   i.v. = intravenous
Note: in this book, reference to penicillin is taken to mean Penicillin G or Penicillin V

Toxicity: virtually non-toxic; very large doses can therefore be given if required. Hypersensitivity is a problem both in the form of rashes (especially with ampicillin) and, rarely, anaphylaxis (with any penicillin given by injection).

Cephalosporins
Chemically similar to the penicillins: the antibacterial activity can be altered by variation in the side chains of the cephalosporin nucleus. A large number of new cephalosporins are now being developed.

Chemical structure:

$$R_1 - \overset{\overset{\displaystyle O}{\|}}{C} - N - R_2$$

[chemical structure diagram of β-lactam ring with S, N, COOH, CH$_2$R$_3$, O]

β -lactam ring

Administration: mostly parenteral although a few oral preparations are available.

Mode of action: bactericidal, similar to penicillin. The cephalosporins are stable to many bacterial β-lactamases and this stability has been increased with the later generations of cephalosporins.

Clinical use: Since most cephalosporins are given parenterally, they are virtually restricted to hospital patients: probably most useful in seriously ill patients especially if infected with more than one organism. Valuable second-line drugs.

Toxicity: low. Rashes, fever, sometimes pain at site of injection. Nephrotoxicity — especially with cephaloridine.

Hypersensitivity: about 20 per cent of people who are hypersensitive to penicillin are also hypersensitive to cephalosporins.

Table 46.3 lists the main cephalosporins, and some of their properties.

Sulphonamides and trimethoprim

Both drugs act sequentially in the synthesis of tetrahydrofolate: widely used in combination because of *in vitro* evidence of synergism: trimethoprim is now available, like sulphonamides, on its own.

Cotrimoxazole

Contains: sulphamethoxazole and trimethoprim in 5 : 1 ratio.

Administration: oral, intramuscular, intravenous.

Mode of action: bacteriostatic: sulphonamide competes with para-amino benzoic acid as a substrate for the enzyme dihydropteroate synthetase which catalyses the synthesis of dihydropteroate which is then converted into dihydrofolate: trimethoprim combines with and blocks dihydrofolate reductase which converts dihydrofolate to tetrahydrofolate: these are sequential steps in the synthesis of methyl-tetrahydrofolate which is required for DNA synthesis.

Table 46.3   The cephalosporins

Cephalosporin	Administration	Antibacterial spectrum	Clinical Use
First generation			
Cephaloridine	i.m., i.v.	Wide range of Gram-positive and Gram-negative bacteria *Strep. faecalis*, *Ps. aeruginosa*, *H. influenzae* and *Bacteroides species* are resistant *Staph. aureus* is sensitive unless methicillin resistant	Second-line drugs: formerly used in severe sepsis; oral drugs may be of value in 'difficult' urinary infections
Cephalothin	(i.m.), i.v.		
Cephalexin	oral		
Cephradine	oral, i.m., i.v.		
Cephazolin	i.m., i.v.		
Second generation			
Cefuroxime	i.m., i.v.	Wide: with marked stability to β-lactamases of Gram-negative as well as Gram-positive bacteria. Active against *H. influenzae* and (especially cefoxitin) *Bacteroides species*. Inferior antistaphylococcal activity	Have largely replaced the first generation drugs in severe systemic infections
Cefoxitin	i.m., i.v.		
Cefamandole	i.m., i.v.		
Third generation			
Cefotaxime	i.m., i.v.	Similar to second-generation drugs but in addition some activity against *Ps. aeruginosa*	Not yet generally available; similar to second-generation drugs

Note: Several more third generation cephalosporins are at an advanced stage of development: many show promise of high-level broad-spectrum activity.

Resistance: is not uncommon: mainly due to production by bacteria of enzymes resistant to the action of sulphonamide and trimethoprim: the resistant enzymes are plasmid-coded and allow normal metabolism despite blockage of the chromosomal enzymes by the drugs: the gene for trimethoprim resistance is often present on a transposon.

Antibacterial spectrum: broad: active against both Gram-positive and Gram-negative bacteria: *Ps. aeruginosa* is resistant.

Clinical use: widely used for urinary and respiratory tract infections: particularly useful in acute exacerbations of chronic bronchitis.

Toxicity: nausea and vomiting: rashes: occasionally thrombocytopenia, leucopaenia: mouth ulceration: folate deficiency has been reported.

Sulphonamides

Now rarely used on their own: still prescribed for urinary tract infection (although acquired resistance is a problem) and, in combination with penicillin, in meningitis because of their excellent penetration into the CSF.

Sulphadimidine is the most widely used preparation.

Trimethoprim

Now available on its own: mainly for urinary tract infections but will probably also be useful in respiratory infections: the main advantage seems to be that the incidence of side effects will be reduced.

Aminoglycosides

A family of extremely useful antibiotics.

Below are the main aminoglycoside antibiotics:

Aminoglycoside	Main clinical use
Gentamicin ⎫ Tobramycin ⎬ Amikacin ⎭	Severe infections in hospital due to coliforms; gentamicin is the most widely used
Streptomycin	Now little used
Kanamycin ⎫ Neomycin ⎭	May be given orally in 'gut sterilisation' regimens prior to surgery, in leukaemia, in chronic liver disease: neomycin also as topical cream

Administration: intramuscular, intravenous.

Mode of action: bactericidal: inhibition of protein synthesis due to action on bacterial ribosomes: all aminoglycosides cause misreading of messenger RNA to produce amino acid substitution in proteins.

Resistance: is generally due to the acquisition of plasmid-coded inactivating enzymes which cause acetylation, adenylation or phosphorylation of the drugs: occasionally due to alteration in the bacterial ribosomes rendering them resistant to the aminoglycoside.

Antibacterial spectrum: coliforms, *Ps. aeruginosa*, staphylococci: streptococci and strict anaerobes are intrinsically resistant.

Toxicity: ototoxicity (vertigo or deafness) and nephrotoxicity are major problems. In patients with renal impairment or on long term therapy, serum levels must be monitored (see p.44): ensure that the trough level is not excessive (2 mg/1 for gentamicin).

Tetracyclines

Amongst the most widely used antibiotics although there are few specific indications for their use: broad spectrum and remarkably free of serious side-effects, they have been particularly successful in the management of chronic bronchitis.

Tetracyclines:
  Tetracycline
  Chlortetracycline
  Oxytetracycline
  Doxycycline ⎫ more effective; can be given in
  Minocycline ⎭ smaller dosage

Administration: almost always oral.

Mode of action: bacteriostatic: inhibit protein synthesis by preventing the attachment of amino acids to ribosomes.

Resistance: is fairly common: generally plasmid-mediated.

Antibacterial spectrum: broad: both Gram-positive and Gram-negative bacteria: however, some strains of *Strep. pyogenes,* pneumococci and *H. influenzae* are now resistant: *Ps. aeruginosa,* and *Proteus species* are intrinsically resistant: active against *Mycoplasma pneumoniae,* rickettsia, *Coxiella burneti* (the cause of Q fever) and chlamydiae.

Main clinical use: acute exacerbations of chronic bronchitis including long-term prophylaxis: treatment of non-specific urethritis, atypical pneumonia, Q fever, psittacosis.

Toxicity: diarrhoea — due to disturbance of alimentary flora — is common but usually mild and self-limiting: rarely there may be renal and liver damage: contraindicated in early pregnancy because the drug may be deposited in the teeth of the developing fetus with permanent yellow staining.

Metronidazole

Originally used — very successfully — to treat *Trichomonas vaginalis* infections in women: now known to be exceedingly effective against anaerobic bacteria.

Administration: oral, rectal (suppositories), intravenous.

Mode of action: bactericidal: converted by anaerobic bacteria to active (reduced) metabolite with inhibitory action on DNA synthesis.

Resistance: almost unknown.

Antibacterial spectrum: strictly anaerobic bacteria e.g. bacteroides, anaerobic cocci and clostridia: no effect on aerobic organisms: actinomyces are resistant: active against protozoal infections (e.g. *Entamoeba histolytica*).

Clinical use: any anaerobic infection, e.g. abdominal and gynaecological wound sepsis, deep abscesses, peritoneal sepsis: Vincent's angina: peroperative prophylaxis in abdominal and gynaecological surgery sometimes by the use of suppositories.

Toxicity: low — sometimes nausea: metallic taste in mouth.

Lincomycin

Clindamycin (7-chloro-7-deoxylincomycin) is an antibiotic active against Gram-positive bacteria and some anaerobes: previously widely used in hospital because it penetrates tissue well and acts on both *Staph. aureus* and *Bacteroides species:* it has recently been the subject of a warning notice from the Committee on the Safety of Medicines because it has been associated more often than other drugs with antibiotic-associated pseudo-membranous colitis: although not banned, it should not be used since suitable alternative drugs are available.

Erythromycin

The most widely used member of the macrolide group of antibiotics.

Administration: oral, intramuscular, intravenous.

Mode of action: bacteriostatic by inhibition of protein synthesis.

Resistance: lavish use of erythromycin in the 1960s led to emergence of resistant *Staph. aureus:* otherwise this has not been a clinical problem.

Antibiotic spectrum: penicillin-like but includes in addition *H. influenzae, Bord. pertussis, Bacteroides species, Campylobacter species, Legionella pneumophila* and *M. pneumoniae.*

Clinical use: staphylococcal infections: a variety of respiratory infections (tonsillitis, sinusitis, bronchitis and pneumonia, diphtheria, whooping cough, Legionnaire's disease): campylobacter enteritis. A useful second-line drug in patients hypersensitive to penicillin.

Toxicity: a safe antibiotic except for one preparation, erythromycin estolate, which may be hepatotoxic.

OTHER ANTIBIOTICS AND ANTIMICROBIAL DRUGS

Some other less commonly used antibiotics and antimicrobial drugs are listed in Table 46.4.

Table 46.4 Some other antimicrobial drugs

Drug	Administration	Mode of action	Antibacterial spectrum	Clinical use	Other features
Fusidic acid	oral i.v.	Bacteriostatic	*Staph. aureus*	Abscesses, osteomyelitis, septicaemia	Penetrates tissues and fluids well
Chloramphenicol	oral i.m., i.v. topical	Bacteriostatic: inhibits protein synthesis	Broad-spectrum	Typhoid fever, *H. influenzae* meningitis, locally eye drops for conjunctivitis	Rarely, causes fatal aplastic anaemia: this has restricted its use
Nalidixic acid	oral	Inhibits DNA replication	Coliforms, not *Ps. aeruginosa*	Urinary infections	Side effects include nausea, rarely visual disturbances
Nitrofurantoin	oral	Unknown	*Strep. faecalis* Coliforms but not *Ps. aeruginosa*	Urinary infections	Nausea, peripheral neuropathy is sometimes seen
Vancomycin	oral i.v.	Bactericidal inhibits cell wall synthesis	Gram-positive bacteria	Antibiotic associated colitis (orally); some cases of septicaemia, infective endocarditis (i.v.)	Nephrotoxic, ototoxic; contra-indicated in renal insufficiency
Polymyxins E (Colistin)	i.m., i.v.	Bactericidal lyses bacteria by affecting permeability of cell membrane	*Ps. aeruginosa;* some other Gram negative bacilli	Urinary and other infections due to *Ps. aeruginosa*	Toxic effects include nephrotoxicity and parasthesiae
B	topical			Topical use only for ear, burns, skin infections	
Spectinomycin	i.m.		*N. gonorrhoeae*	Penicillin-resistant gonorrhoea	

ANTITUBERCULOUS CHEMOTHERAPY

For more than 20 years, tuberculosis has been successfully treated with a combination of streptomycin, para-amino salicylic acid (PAS) and isoniazid. Problems with streptomycin toxicity, the fact that it has to be given by injection and the unpalatability of PAS have caused this regimen to be replaced.

Combination of antimicrobial drugs is essential in tuberculosis to prevent the emergence of resistant bacteria: this was frequent in early days of chemotherapy when single drugs were given.

Antituberculous drugs in current use
1. **Isoniazid** (isonicotinyl hydrazide)
   Administration: oral, parenteral preparations available for instillation.
   Mode of action: bacteriostatic, penetrates well into tissues and fluids and acts on intracellular organisms.
   Resistance: develops readily.
   Toxicity: uncommon, peripheral neuritis, psychotic and epileptic episodes.

2. **Rifampicin**
   Administration: oral.
   Mode of action: inhibits by combining with bacterial DNA-dependent RNA polymerase.
   Resistance: develops rapidly unless other drugs are used in combination.
   Toxicity: low, liver function may be affected: often transient hypersensitivity: rarely thrombocytopenia.
   Contraindicated: in the first trimester of pregnancy.
   Patients should be warned that urine, sputum, tears, become coloured red as a result of the therapy.

3. **Ethambutol**
   Administration: oral.
   Resistance: uncommon.
   Toxicity: optic neuritis may develop: it is generally reversible and is uncommon if low dosages are used.

Other (second-line) drugs which may be used:
Streptomycin
Pyrazinamide
Prothionamide
Thiacetazone
Cycloserine
Capreomycin

Recommended schedule for pulmonary tuberculosis

Without cavitation: 6 months course of isoniazid, rifampicin and ethambutol: ethambutol is stopped after 8 weeks.

With cavitation: treatment should be continued for 9 months. Three months courses of treatment are under trial at present.

ANTIBIOTIC RESISTANCE IN BACTERIA

A major problem in antibiotic therapy is the emergence of drug-resistant bacteria. The frequency of this depends on the organism and the antibiotic concerned. Some organisms rapidly acquire resistance, e.g. certain coliforms, *Staph. aureus:* others rarely do so, e.g. *Strep. pyogenes.* Resistance to some antibiotics virtually never develops, e.g. metronidazole, whereas with others resistant strains readily emerge, e.g. penicillin, tetracycline, streptomycin.

Drug resistance in clinical practice is associated with antibiotic use: when a small number of resistant bacteria have emerged they will be at a selective advantage in the presence of the antibiotic and will multiply at the expense of sensitive bacteria. The widespread, often indiscriminate prescribing of antibiotics in hospitals has therefore favoured the survival and increase of drug-resistant bacteria.

Drug resistance is of three types:

1. **Primary resistance**
   An innate property of the bacterium and unrelated to contact with the drug, e.g. the resistance of *Esch. coli* to penicillin.

2. **Acquired resistance**
   Due to *mutation* or *gene transfer.*
   Mutation: is when resistant bacteria arise as a result of spontaneous mutation: this may be relatively infrequent or — in the case of *Myco. tuberculosis* and streptomycin — common.
   Gene transfer — a major cause of resistance in bacteria — enables resistance to spread from bacterium to bacterium not only within a species but crossing between different species in the case of many coliforms. See Chapter 4.
   Cross resistance. This is when resistance to one antimicrobial drug confers resistance to other — usually chemically related — drugs, e.g. bacteria resistant to one tetracycline or one sulphonamide are resistant to all tetracyclines or all sulphonamides respectively.
   Conversely, *dissociated resistance* is also seen when resistance to one drug is not accompanied by resistance to closely related drugs, e.g. resistance to gentamicin is not always associated with resistance to tobramycin.

Mechanisms of antibiotic resistance

There are three main mechanisms:

1. *Permeability:* the cell membrane becomes altered — possibly by modification of proteins in the outer cell membrane — so that antibiotics or other antimicrobial drugs cannot enter and be taken up by the bacterial cell: often associated with low level resistance to several drugs, but occasionally with high level resistance to a single drug (e.g. tetracycline): a common type of antibiotic resistance in *Ps. aeruginosa.*

2. *Modification of the site of action* of the antimicrobial drug in which the enzyme or substrate with which the drug reacts becomes resistant to it and is able to function normally in the presence of the drug, e.g. trimethoprim and sulphonamide resistance: in the case of trimethoprim resistance, a plasmid or transposon coding for a resistant dihydrofolate reductase is acquired by the bacterium — the bacterium retaining its own chromosomal-coded trimethoprim-sensitive enzyme.

3. *Inactivation of the antibiotic:* is a common mechanism of resistance: the antibiotic is inactivated by enzymes produced by the bacterium, again for the most part plasmid-coded: e.g. β-lactamase destruction of the β-lactam ring responsible for the antibacterial action of penicillins: acetylating, adenylating and phosphorylating enzymes in the case of the aminoglycosides.

PRINCIPLES OF ANTIMICROBIAL THERAPY

1. *Administration* is indicated for an established infection that makes a patient sufficiently ill to require specific treatment: trivial, self-limiting infections in healthy individuals should not be treated with antibiotics.

2. *Choice of drug:* successful chemotherapy depends on the infecting organism being sensitive to the drug chosen. Attempts to treat *viral,* respiratory infections — a common practice — are doomed to failure unless there is secondary *bacterial* infection.

   The choice of drug is based on:

   a. *Clinical diagnosis:* implies prescribing on an informed best-guess basis: most infections requiring antibiotics have in fact to be treated before laboratory results are available: they vary in severity from exacerbations of chronic bronchitis to septicaemia. Sometimes the clinical diagnosis indicates a specific bacterial cause, e.g. boil, typhoid fever, but often the cause cannot be deduced from the clinical picture, e.g. peritonitis, urinary infection.

All medical students must therefore have a working knowledge of the bacteriology of infection so that they can prescribe effectively.

b. *Laboratory diagnosis:* adequate specimens for diagnosis should, whenever possible, be taken before chemotherapy is started. Isolation of the pathogen and sensitivity tests take time — 24 h at least and usually longer — and as soon as results are available, treatment should be reviewed. Laboratory monitoring can make the difference between success and failure of therapy.

3. *Route of administration:* drugs must be given parenterally to seriously ill patients. Oral antibiotics must be both acid stable (e.g. penicillin V is acid stable but penicillin G is not) and absorbed from the gastrointestinal tract.

4. *Dosage:* must be adequate to produce a concentration of antibiotic at the site of infection greater than that required to inhibit the growth of the infecting organism.

5. *Duration:* treatment should continue until eradication or until the body defence mechanisms are able to cope with the infection.

6. *Distribution:* the drug must penetrate to the site of infection, e.g. in meningitis the antibiotic must pass into the CSF. Deep-seated sepsis is a particular problem and an important cause of antibiotic failure: antibiotics cannot penetrate 'walled-off' abscesses or internal collections of pus, and treatment will probably fail unless the pus is drained. Surgical intervention is also necessary if there are established pathological changes, e.g. urinary obstruction due to stones, chronic tuberculous cavities.

7. *Excretion:* agents used to treat urinary infections is excreted in the urine in high concentrations; some, e.g. nalidixic acid and nitrofurantoin, do not achieve useful serum levels but are eliminated almost exclusively by the renal route and are therefore indicated only for the management of urinary infections.

Urinary pH affects the activity of some drugs, e.g. the aminoglycosides are far more active in an alkaline medium; the reverse is true of nitrofurantoin, which should not therefore be used to treat infections caused by *Proteus* species which raise the pH of the urine.

Erythromycin is excreted largely in the bile; only low concentrations can be detected in urine.

8. *Toxicity:* although the antibiotics in general use are well-tested safe drugs, patients should be warned of possible side effects, e.g. the mild diarrhoea that is common with tetracycline therapy or the red colouring of body fluids with rifampicin.

Serious toxicity can manifest itself in two ways:

a. *Direct toxicity:* e.g. ototoxicity with the aminoglycosides; nephrotoxicity with the polymyxins and the rare bone marrow aplasia due to chloramphenicol

b. *Hypersensitivity* is most often due to the penicillins

Other complications include *superinfection* with antibiotic-resistant microorganisms, e.g. coliforms and yeasts: this is surprisingly uncommon in immunologically normal patients, e.g. chronic bronchitics on repeated courses of tetracycline, but is a major problem in compromised hosts. Antibiotic-associated pseudo-membranous colitis is probably a special example of this.

9. *Use of drugs in combination:* two advantages have been claimed for this:

a. *Emergence of drug-resistant bacteria will be prevented:* certainly true in the treatment of tuberculosis and may apply in infections due to *Staph. aureus* — an organism with a marked propensity to become resistant — especially when an antibiotic is being given to which resistant strains can emerge during clinical therapy (e.g. fusidic acid).

b. *Enhanced antibacterial effect (synergism) will be achieved:* this is governed to some extent by the *'Jawetz Law'* on combined action although there are exceptions. It depends on whether each component is bacteriostatic or bactericidal in action and predicts that the effect of a combination will be as follows:

Bactericidal + bactericidal: may be synergistic
Bactericidal + bacteriostatic: may be antagonistic
Bacteriostatic + bacteriostatic: will be additive

This working rule, however, has some value in clinical practice. Combinations of a penicillin and an aminoglycoside, both bactericidal drugs, are often synergistic and may be necessary to eradicate infection in endocarditis. There was disastrous antagonism between penicillin and tetracycline, a bactericidal and a bacteriostatic drug, when these were used in the treatment of pneumococcal meningitis.

10. *Antibiotic prophylaxis:* early but indiscriminate attempts to prevent infection by giving antibiotics for several days or more failed. This was because the infection to be avoided, e.g. pneumonia in unconscious patients or wound infection after surgery, could be due to a number of different bacteria not all of which are sensitive to the drugs chosen: prolonged antibiotic administration therefore resulted in the selection of resistant organisms which subsequently caused infection. Successful

prophylaxis is possible only when the pathogens are always, or almost always, sensitive to the drug or drugs prescribed.

Short-term prophylaxis: one or at most a few doses of carefully chosen antibiotics given to cover the time when the risk of an infection being established is greatest — i.e. usually the peroperative period. This is a controversial issue. Some examples are listed in Table 46.5.

Table 46.5 Antibiotic prophylaxis

Clinical situation	Bacteria most likely to cause infection	Appropriate prophylactic regimen
Colonic surgery	Coliforms anaerobes (bacteroides, anaerobic cocci, clostridia)	Gentamicin with metronidazole
Appendicectomy	Anaerobes (as above)	Metronidazole
Gynaecological surgery	Anaerobes (as above)	Metronidazole
Biliary tract surgery	Coliforms	Cephalosporin or cotrimoxazole
Insertion of prosthetic heart valves	*Staph. aureus, Staph. epidermidis,* corynebacteria	Cephalosporin or cloxacillin with gentamicin
Insertion of prosthetic joints	*Staph. aureus, Staph. epidermidis,* corynebacteria	Cloxacillin alone or with gentamicin : fusidic acid
Amputation of ischaemic limb	*Cl. welchii*	Penicillin
Dental extraction in patients with heart valve disease	Oral streptococci	Penicillin alone or with streptomycin : Amoxycillin
Prevention of tetanus* after wounding	*Cl. tetani*	Penicillin

*In conjunction with immunoprophylaxis.

Long-term prophylaxis: may be continued for months or years.

Rheumatic fever: recurrence of rheumatic fever invariably follows a throat infection with *Strep. pyogenes,* which is always penicillin sensitive. The incidence of further attacks is greatly reduced by giving penicillin.

Urinary tract infection: may be avoided in women who suffer repeated episodes by administration of cotrimoxazole or nitrofurantoin since some 90 per cent of urinary pathogens are sensitive to these drugs.

11. *'Chemotherapy without bacteriology is guesswork':* collaboration between clinician and bacteriologist is crucial especially in the severe infections increasingly encountered in hospitals today. Frequent discussions between ward and laboratory are to be encouraged; they are of benefit to both.

47

Prophylactic immunisation

The introduction of immunisation against infectious disease has been one of the most successful developments in medicine.

Immunisation aims to produce immunity to a disease artificially and without ill-effects.

Immunity can be classified under two main headings:

IMMUNITY

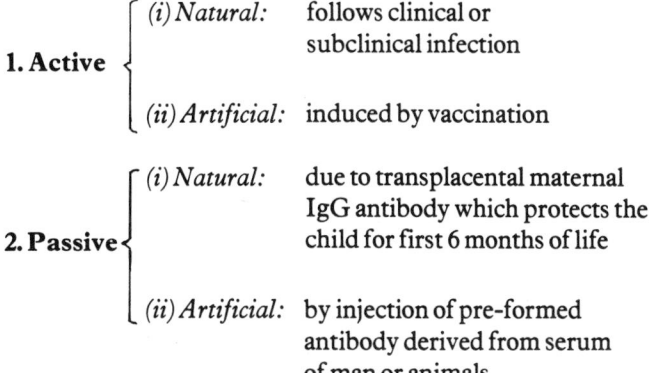

1. Active	*(i) Natural:*	follows clinical or subclinical infection
	(ii) Artificial:	induced by vaccination
2. Passive	*(i) Natural:*	due to transplacental maternal IgG antibody which protects the child for first 6 months of life
	(ii) Artificial:	by injection of pre-formed antibody derived from serum of man or animals

Artificial immunisation can therefore be used to produce both active and passive immunity:

1. Active immunity

By administration of vaccine to provoke an immune response with production of antibody; sometimes cell-mediated immunity is also produced.

- a. *Antibody:* protects in different ways depending on the type of disease; most effective in virus diseases because antibody neutralises virus infectivity. In bacterial disease due to exotoxin, antibody neutralises the toxin; in both bacterial and virus infections antibody enhances phagocytosis.
- b. *Cell mediated immunity:* is stimulated independently of antibody: particularly important in resistance to chronic

bacterial infections characterised by intracellular parasitism (e.g. tuberculosis, leprosy, brucellosis).

The onset of immunity is delayed but when established lasts for years, sometimes for life.

2. Passive immunity

By injection of pre-formed antibody present in human or animal serum: immediate immunity is conferred but it is short-lasting — usually for only a matter of weeks.

Figure 47.1 shows antibody levels after passive immunisation and active immunisation with live attenuated and killed organisms.

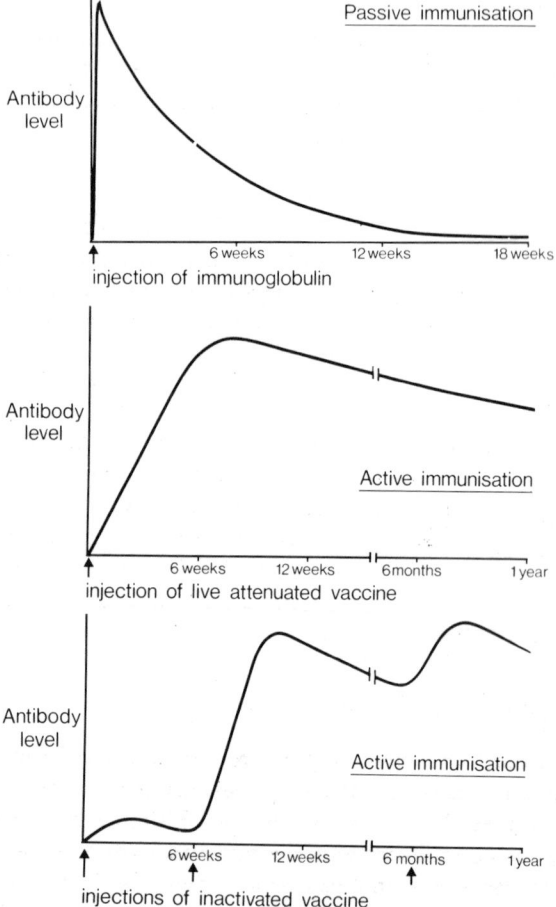

Fig. 47.1 Antibody levels following different methods of immunisation.

ACTIVE IMMUNISATION

Objective: is to administer an antigenic preparation or vaccine to produce immunity based on adequate antibody levels and a population of cells with immunological memory.

Long-lasting: the specific immunity once produced persists and even after many years infection may stimulate an accelerated antibody response.

VACCINES

Vaccines are of three types:

1. *Live attenuated organisms*
Multiply in the body and mimic natural infection with antibody production but without symptoms: mild reactions are similar to the natural disease; a single dose gives long-lasting immunity which can be reinforced with later booster doses.

2. *Killed (inactivated) organisms*
Several — usually three — doses are required because there is no multiplication in the body: as a general rule the first two 6 weeks apart, the second and third 6 months apart; later booster doses are also necessary; reactions do not resemble those of the natural disease and usually follow soon after inoculation.

3. *Toxoids or modified toxic products*
Very successful vaccines in diseases due to a single exotoxin, e.g. diphtheria, tetanus: toxoids are toxins rendered harmless — usually by formaldehyde — but retaining antigenicity. Antigenicity of toxoids can be increased by absorption to a mineral carrier (such as aluminium salts) or when mixed with a suspension of lipopolysaccharide endotoxin, e.g. contained in the pertussis component of triple vaccine (pertussis, diphtheria, tetanus).

ASSESSMENT OF PROTECTION

Immunisation is of value only if it results in a significant degree of protection against infection: live-attenuated organisms and toxoids are better vaccines than killed organisms. Widespread vaccination of human populations — the aim is usually to achieve acceptance rates of 70 per cent or over — has caused a dramatic fall in the incidence of many infectious diseases, e.g. diphtheria, poliomyelitis: however, in some instances when the introduction of a vaccine has coincided with other measures to control the spread of infection (e.g. improvements in housing, sanitation, nutrition or the use of chemotherapy which reduced the duration of infectivity) these may have had more effect on the incidence of the disease than vaccination, e.g. tuberculosis.

Field trials, in which the effect of a vaccine in an at-risk population is studied, are of crucial importance: *observe for* reduction in the attack rate or the severity of the disease in vaccinees. Extensive carefully controlled field trials in human populations are essential in the final evaluation of most vaccines.

Antibody levels (titres) after vaccination and occasionally the appearance of cell-mediated immunity (e.g. tuberculin conversion after BCG vaccination) are determined as a routine when a vaccine is being developed. Detailed studies in human volunteers are easy to organise and the results give some indication about the likelihood of the vaccine providing worthwhile protection. However, the ability to stimulate antibody production does not guarantee effective prophylaxis. The reasons for the failure of an apparently promising vaccine are complex: in some instances it may be that an inappropriate antibody is formed, e.g. IgM and IgG antibody but not secretory IgA are formed after parenteral administration of killed vaccine although secretory IgA antibody is the principal protective antibody at mucosal surfaces and is therefore of great importance in respiratory virus infections (e.g. influenza) and also in poliomyelitis and cholera (where the gut is the site of primary multiplication).

CONTROL OF VACCINE PREPARATION

Stringent quality control in the manufacture of each batch of an established vaccine is essential to ensure continued safety and potency.

Safety: the following are the main problems: fortunately now all are very rare.

1. Contamination: usually of attenuated viral vaccines with other live viruses from the tissue culture used for virus propagation.
2. Inadequate inactivation of killed vaccine.
3. Reversion to virulence of attenuated vaccine.
4. Residual toxicity of toxoids.

Potency: this is ensured by the following measures:

1. Live vaccines are prepared so that each batch is only a limited number of subcultures away from a known parent strain.
2. Killed vaccines must be prepared from a strain of organism containing all the antigens involved in the protective immune response.

Tests of potency usually depend on measurement of the antibodies produced in response to inoculation of the vaccine into experimental animals: other tests assess the survival of immunised animals after challenge with virulent organisms but the method of the test may bear little relationship to the human disease e.g. assay of pertussis vaccine is by testing for protection of mice against intracerebral injection with *Bordetella pertussis* and this is said to correlate with the ability of the vaccine to prevent human disease.

ADMINISTRATION

1. *Age*

Immunisation must aim at those at greatest risk: some vaccines (e.g. typhoid and cholera) are indicated regardless of age for anyone entering an endemic area: others (e.g. influenza) are given to those, usually middle-aged or elderly, with chronic cardiac or respiratory disease.

Childhood: most vaccines, however, are given to children because most of the diseases they prevent are mainly encountered in childhood: e.g. more than two-thirds of the deaths from pertussis are in infants under one year old.

There are two problems in starting immunisation in the first months of life.

a. *The infant immune system is not fully developed* at birth and capacity to make antibody is therefore limited: nevertheless it is probably adequate if the vaccine is potent.

b. *Transplacental maternal antibody* may prevent a response to live virus vaccines and reduce that to some killed vaccines.

Official policy in Britain is to start immunisation at 3 months old: this is a compromise; delay until the child is 6 to 9 months old would produce better responses but the chance of establishing immunity when it is most needed would be lost.

2. *Combined vaccines*

Giving more than one vaccine at a time is attractive because it reduces the number of injections required and therefore increases acceptability by the parent and child.

Combined vaccines may enhance antibody production (e.g. the presence of the pertussis component acts as an adjuvant for the toxoids in triple vaccine): but sometimes the response to one organism diminishes that to others; the effect of combining vaccines can only be determined after much trial and research.

3. *Live vaccines*

These must generally be given 3 weeks apart: if not, the risk of reactions is sometimes increased; there may also be interference so that multiplication of one of the viruses is suppressed, preventing an immune response.

COMPLICATIONS OF IMMUNISATION

Side effects: are common after administration of some vaccines, e.g. many killed bacterial vaccines cause local (pain and redness at the injection site) and general (fever and constitutional upset) reactions; although unpleasant these are usually trivial.

Serious complications: although rare, are associated with certain vaccines: these mainly affect the CNS and can result in permanent brain damage.

The small but definite risk of serious reactions in a tiny proportion of vaccinees must be balanced against the benefits to the larger number protected: the present controversy over pertussis vaccine illustrates the difficulty that this may present.

Vaccination is contraindicated in the following circumstances:

1. If there has been a previous severe local or generalised reaction to that vaccine.
2. Live vaccines should never be given to
   a. *Immunocompromised patients* — because of the risk of severe generalised infections.
   b. *Pregnant women* — because of the danger of transplacental spread to the fetus.

VACCINES IN CURRENT USE

Below is a comprehensive list of bacterial vaccines; also included are the main virus vaccines in current use.

LIVE VACCINES

BCG vaccine

Contains: live *Mycobacterium bovis* attenuated by propagation in a bile-potato medium (Bacille Calmette-Guérin). Killed vaccines are of no value as immunising agents in tuberculosis since they do not produce a cell mediated response.

Indications: policy in the UK is to vaccinate all children between 10 and 13 years of age *after* a tuberculin test has shown that they are non-reactors. Give at birth to infants at high risk of contact with tuberculosis, e.g. those known to have a relative with the disease.

Table 47.1 Vaccines in current use

	Bacterial vaccines	Viral vaccines
Live	BCG (tuberculosis)	Poliomyelitis (Sabin) Measles Mumps Rubella Smallpox (vaccinia)
Killed	Typhoid and paratyphoid Cholera Plague Pertussis Pneumococcal extract	Influenza Poliomyelitis (Salk)
Toxoids	Diphtheria Tetanus	

Administration: one dose intradermally over the deltoid area. Normally a red papule develops at the site of injection after some weeks and soon subsides.

Adverse reactions: the papule may progress to an indolent ulcer and discharge pus: associated axillary lymphadenopathy may develop.

Protection: MRC field trials in the UK (1950-71) and studies in North America both showed durable (10–15 year) protection. The incidence of clinical disease in vaccinees was reduced by 80 per cent.

However, other field trials have yielded less encouraging results.

Poliomyelitis vaccine

Contains: live-attenuated strains of polio virus types 1, 2 and 3 — Sabin vaccine. (Salk vaccine which was developed earlier contains inactivated strains of polio virus types 1, 2 and 3: it is no longer in routine use in Britain.)

Indications: active immunisation of all infants starting when 3 to 6 months old.

Administration: orally: three spaced doses are required to ensure multiplication in gut with both local — i.e. gut — and serum antibody production to each of the three types. Booster doses are recommended at school entry and on leaving school.

(Salk vaccine is given by injection and produces serum antibodies only.)

Adverse reactions: minimal: rare cases of paralysis in adults due to Sabin type 3 virus.

Protection: excellent with both vaccines. Following a vigorous campaign with Salk vaccine in the UK launched in 1956 there was a

100-fold decrease in poliomyelitis notifications between 1957 and 1963 and an 80 per cent protection against paralytic disease. Sabin vaccine introduced later has been equally effective.

Sabin vaccine is preferred because:

1. It produces gut immunity by stimulating the formation of secretory IgA (copro or alimentary antibody) — killed vaccine cannot achieve this. As a result wild virulent virus cannot find enough susceptible hosts in which to circulate and dies out.
2. Smaller quantities of virus are required.
3. It is easier to administer and can be given more quickly in the face of an epidemic.

Measles vaccine

Contains: live attenuated virus.

Indications: active immunisation of all children in the second year of life to prevent the respiratory complication of measles and the mental retardation that may follow measles encephalitis.

Although the disease is usually mild in the UK, in some developing countries measles can still be very severe — mortality rates around 10 per cent have been recorded recently in tropical Africa. The acceptance rate of vaccination in the UK is now only about 50 per cent.

Administration: one dose by injection.

Adverse reactions: few, the present vaccine is well-tolerated: fever and transient rash — like mild measles — may follow 6–12 days after vaccination. Post infectious encephalomyelitis, a serious complication, is rare — about 1 case per million.

Protection: good and apparently long-lasting; probably slightly less solid than that following natural infection; so far there is no evidence that immunity deteriorates significantly with increasing age. It would be serious if childhood immunity were to wane and leave some people susceptible in adult life.

Mumps vaccine

Contains: live attenuated virus.

Indications: prevention of mumps in children over one-year old and in adults.

Administration: one dose by injection.

Adverse reactions: few; occasionally fever, very rarely parotitis.

Protection: substantial durable immunity for 10 years at least, probably for much longer.

Note: This vaccine is widely used in the USA alone or in combination with measles and/or rubella vaccine.

Rubella vaccine

Contains: live attenuated virus.

Indications: in the UK the vaccine is administered to all girls between 11 and 14 years of age and to susceptible adult women post-partum. This policy, aimed at preventing congenital rubella, does not affect the epidemiology of the disease and outbreaks of rubella have continued to take place in which some susceptible pregnant women have contracted infection.

Note: A past history of rubella unconfirmed by laboratory tests is an unreliable guide to immune status: other virus diseases can mimic rubella clinically.

Administration: one dose by injection.

Adverse reactions: uncommon but there may be mild rubella-like symptoms including arthralgia some 9 days after vaccination. Pregnancy must be avoided for 2 months after vaccination: the vaccine must never be given during pregnancy.

Protection: apparently good with long-lasting immunity.

Smallpox vaccine

Contains: live vaccinia virus.

Indications: few, now that smallpox has been eradicated. Routine immunisation in childhood has not been advocated in the UK since 1971 and international certificates are rarely required although 23 tropical countries still insisted on evidence of vaccination in January 1980.

Administration: by multiple light pressures with a needle through a drop of vaccine placed on the skin of the upper arm. The 'take' may be localised primary vaccinia, revaccination (milder) vaccinia or an accelerated response (in the partially immune). When there is no vesiculation the vaccination should be repeated.

Adverse reactions: rare, usually associated with primary vaccination. The most severe are generalised vaccinia and post-infectious encephalomyelitis.

Protection: solid immunity for at least 3 years.

KILLED VACCINES

Typhoid and paratyphoid (TAB) vaccine

Contains: heat-killed phenol-preserved suspension of *Salmonella typhi* and *S. paratyphi A and B*.

Indications: for those travelling to or living in areas where enteric fever is endemic.

Administration: two spaced doses by injection: booster doses every 3 years.

Adverse reactions: local and general reactions are common: early in onset, they subside in 24–48 hours.

Protection: apparently better against typhoid than the paratyphoid fevers. Extensive field trials in Yugoslavia (1954–55) carried out by WHO showed a protection rate against typhoid of 70 per cent. An alcohol-killed alcohol-preserved vaccine (alcohol preserves the Vi — 'virulence' — antigen) evaluated at the same time was ineffective.

Cholera vaccine

Contains: heat-killed *Vibrio cholerae* serotypes Inaba and Ogawa.

Indications: for those travelling to and resident in endemic areas; its use on a large scale has been advocated during epidemics.

Administration: two spaced doses by injection; booster doses every 6 months.

Adverse reactions: local and general reactions are quite common; serious reactions are rare.

Protection: poor, estimated at 50 per cent, and short-lasting — about 6 months: there is a similar degree of cross-protection to the El Tor biotype which is antigenically identical and which is at present pandemic.

Attempts have been made to produce better vaccines, e.g. live avirulent oral vaccines to stimulate secretory IgA antibodies in the gut (coproantibodies) as well as serum antibodies and a toxoid prepared from cholera enterotoxin (exotoxin): none has so far gained acceptance.

Plague vaccine

Contains: formaldehyde-killed *Yersinia pestis.*

Indications: as a prophylactic where plague is endemic: during epidemics.

Administration: three spaced doses by injection; revaccination necessary every 6–12 months.

Adverse reactions: mild local reactions are common: with repeated doses there may be severe generalised reactions.

Protection: uncertain and short-lived: said to reduce the severity of the disease.

Pertussis vaccine

Contains: killed, freshly-isolated, smooth strains of *Bordetella pertussis.* The vaccine should contain all the surface antigens of *Bord. pertussis* associated with epidemics: these antigens designate the three common serotypes — 1,3; 1,2,3; and 1,2. However, it is still not certain if these antigens are responsible for a protective response. Manufacturers select their own strains for vaccine production and there are significant differences in the protection afforded by individual vaccines and in the incidence of adverse reactions they cause.

Indications: official policy in Britain is the active immunisation of all children starting at 3–6 months old. The heated debate which began in 1974 about the risk and efficacy of vaccination aroused much public concern. As a result, the acceptance rate has declined from 75 per cent to less than 40 per cent (unfortunately also reducing the acceptance rate of diphtheria and tetanus immunisation because all three are contained in Triple Vaccine).

Administration: three spaced injections: booster doses are not recommended because pertussis is not a problem after 5 years of age.

Adverse reactions: usually local and trivial: one infant in 40 develops excessive crying and irritability. Severe reactions attributed to the vaccine are convulsions (1 in 5000) and permanent brain damage with mental retardation (1 in 35 000). However, it has been argued by supporters of the vaccine that these complications are coincidental and not causally related to vaccination.

Protection: many early vaccines were useless and the MRC conducted extensive field trials (1951–59) to examine the reduction of attack rate in vaccinated home contacts: although the poorest vaccines were of no value, the best gave impressive protection and, in general, when disease developed, it was modified. As a result, mass vaccination was started in the UK. During the following 20 years there have been conflicting claims about the efficacy of different vaccines. Whooping cough is a disease that has become less severe as living conditions have improved and children have become healthier: but it has not been eliminated.

Pneumococcal vaccine

Contains: a saline solution of 14 highly purified *capsular polysaccharides* extracted from pneumococci of the most prevalent pathogenic types, i.e. 1,2,3,4,6,8,9,12,14,19,23,25,51 and 56.

Indications: to prevent pneumococcal pneumonia, bacteraemia and meningitis in individuals at special risk.

Administration; one dose by injection.

Adverse reactions: local in about half and fever in 10 per cent of those vaccinated.

Protection: apparently good against infections caused by the serotypes present in the vaccine.

Influenza vaccine

Contains: inactivated virus: two, sometimes three of the currently circulating strains of influenza A virus; current influenza B strain also included; the virus is propagated in the chick embryo.

Indications:

1. Those likely to suffer severe illness, e.g. elderly people with pre-existing cardio-respiratory disease.
2. Key personnel in essential services, e.g. hospital staff, the police force.

Administration: one dose by injection. Vaccination requires to be repeated each winter.

Adverse reactions: few and mild but a number of cases of polyneuritis (Guillain-Barré syndrome), some severe, were recorded in USA following mass vaccination in 1976-77; there may be severe reactions in those hypersensitive to egg protein.

Protection: short-lived, a matter of a few months; at best the protection conferred is of the order of 60 per cent.

A live-attenuated vaccine administered intranasally has had trials in Eastern Europe; it stimulates the production of secretory IgA on the respiratory mucous membranes but causes more side effects and has not gained acceptance.

TOXOIDS

Diphtheria toxoid

Contains: diphtheria formol toxoid (toxin treated with formaldehyde) — available also absorbed onto aluminium salts.

Indications: children and selected 'at-risk' adults, e.g. hospital or laboratory staff.

Administration: three spaced injections starting at 3–6 months old — usually as part of the Triple Vaccine. Booster dose at school entry.

Adverse reactions: mild and transient under 10 years of age: older children and adults may experience severe side effects and in this group a preliminary Schick test is advised: toxoid is given only to those who are positive reactors; pseudo reactors should not be immunised.

Protection: excellent — it is generally agreed that the disappearance of diphtheria in the UK between 1941 and 1951 was due to immunisation and the disease is now extremely rare in this country. However, in the USA there are still 100–200 cases per year primarily in inadequately immunised infants and children in lower socio-economic groups.

Tetanus toxoid

Contains: tetanus formol toxoid (toxin treated with formaldehyde) — available also absorbed onto aluminium salts.

Indications: the aim is active immunisation of the entire population; *note* that tetanus may develop after common, trivial wounds.

Administration: three spaced injections starting in infancy — usually as part of the Triple Vaccine. Otherwise the course should begin when a situation of risk presents, e.g. after injury at the casualty department. Booster doses every 5–10 years and in the event of injury.

Adverse reactions: rare and mild; severe reactions restricted to adults, usually those who have been hyperimmunised — i.e. who have had too many booster injections.

Protection: excellent.

Triple vaccine

Contains: killed *Bordetella pertussis,* diphtheria toxoid and tetanus toxoid.

Indications: active immunisation of all infants.

Administration: three spaced doses by injection. At school entry, booster doses of diphtheria and tetanus toxoids only.

Adverse reactions: see individual vaccines.

Protection: see individual vaccines.

The scheme for active immunisation officially recommended in the United Kingdom is shown in Table 47.2.

PASSIVE IMMUNISATION

Objective is to produce immunity immediately by the injection of antibodies present in human or animal serum. These antisera are also used in treatment.

Short-lasting: the immunity that follows wanes in a matter of weeks or a few months.

SPECIFIC IMMUNOGLOBULINS

1. HUMAN:

Prepared from plasma pools containing high levels of the appropriate antibody. Donations are taken from individuals after their plasma has been screened for antibody content: sometimes they are chosen because they are recovering from the infection (convalescent serum) or have recently been actively immunised against the disease. Human immunoglobulins have a half life of 26 days after injection and significant protection may last up to 3 months or sometimes longer.

Table 47.2 Schedule of vaccination and immunisation recommended in the United Kingdom

Age	Vaccine	Notes
During the first year of life	Triple Vaccine and oral polio vaccine. Triple Vaccine and oral polio vaccine Triple Vaccine and oral polio vaccine	6-8 weeks between the 1st and 2nd doses; 4-6 months between the 2nd and 3rd doses. First dose should be given between 3 and 6 months
During the second year of life	Measles vaccine	
5 years (school entry)	Diphtheria-tetanus toxoid Oral polio vaccine	Booster dose Booster dose
Between 10-13 years	BCG vaccine	For tuberculin-negative children
Between 11-14 years	Rubella vaccine	Administer to *all girls* irrespective of past history of an attack of rubella
Between 15-19 years (school leaving)	Oral polio vaccine Tetanus toxoid	Booster dose Booster dose

Preparations

Some of the available preparations are listed below:

Hepatitis B immunoglobulin: Indication: post-exposure prophylaxis after accidental inoculation with HBsAg positive blood.

Tetanus immunoglobulin: Indication: prophylaxis and treatment of tetanus.

Varicella-zoster immunoglobulin: Indication: treatment of the disease in immunocompromised hosts who may develop severe varicella or generalised zoster.

Diphtheria immunoglobulin: Indication: prophylaxis and treatment of diphtheria.

Rabies immunoglobulin: Indication: post-exposure prophylaxis: given in association with active immunisation.

2. ANIMAL

Specific antibodies, raised in horses by active immunisation, were used extensively in the past. Unfortunately they contain foreign protein to which the recipient forms antibody with the result that:

1. They are rapidly eliminated — much faster than human immunoglobulins.
2. They can cause unpleasant and sometimes dangerous hypersensitivity reactions including anaphylaxis.

Equine antisera still available include diphtheria antitoxin and polyvalent gas-gangrene antitoxin (of doubtful value as a therapeutic agent).

NON-SPECIFIC IMMUNOGLOBULINS

Human immunoglobulin prepared from donations of pooled normal plasma: contains antibodies to a wide range of infective agents likely to have been encountered by most people.

Indications:

1. *Prophylaxis of infective hepatitis* (Hepatitis A). Administer to travellers going to endemic areas: family contacts of a case.
2. *Prophylaxis of measles:* given within 6 days of exposure will prevent or modify the disease.
3. *To boost immunoglobulin levels* in children with hypogammaglobulinaemia.

Recommended reading

Benenson, A. S. (Ed.) (1975) *Control of Communicable Diseases in Man.*
12th edn. Washington: American Public Health Association.
Christie, A. B. (1981) *Infectious Diseases.* 3rd edn. Edinburgh:
Churchill Livingstone.
Wilson, G. S. & Miles, A. (1975) *Topley and Wilson's Principles of
Bacteriology, Virology and Immunity* Vols. 1 & 2. 6th edn. London:
Edward Arnold.

Index

Abscesses, 230-233
  clinical features, 232
  periapical, 282
  treatment, 233
Abortion in brucellosis, 295
  septic, 225
Acholeplasma, 131, 132
Acid-fast bacilli, 103
Acinetobacter, 83
Acne vulgaris, 102, 220
Actinobacillus, 96
Actinomyces, 106
Actinomycosis, 284
Active immunisation, 331
  immunity, 329
Acute follicular tonsillitis, 161
  glomerulonephritis, 163
Aeromonas, 83
Agar, 38
Agglutination, 45
Aggressins, 149
Aminoglycosides, 318
Anaerobes, *Clostridium*, 115-119
  cocci, 123
  non-sporing, 120-123
Angular cheilitis, 285
Animal anthrax, 297
  brucellosis, 295
Animals as source of infection, 152
Antibiotic sensitivity of *Bacteroides*, 122
  Bordetella, 95
  Brucella, 93
  Campylobacter, 88
  Candida, 135
  Clostridium, 116
  enterobacteria, 71-79
  gonococcus, 110
  Haemophilus, 92
  Legionella, 126
  meningococcus, 111
  mycobacteria, 104

Mycoplasma, 132
Pseudomonas, 82
spirochaetes, 128
Staphylococcus, 60
Streptococcus, 66, 70
tests, 43-45
Vibrio, 87
Antibiotics, 313-328
  aminoglycosides, 318
  and normal flora, 139
  associated colitis, 194
  cephalosporins, 315
  erythromycin, 320
  lincomycin, 320
  metronidazole, 319
  penicillins, 314
  prophylaxis, 326
  resistance, 323
  tetracyclines, 319
  therapeutic principles, 324-328
    in endocarditis, 244, 245
Antibody in active immunity, 329
  maternal, 333
  response, 146
Antigens, 146
  in Gram-negative bacteria, 73
  in identification, 42
Antimicrobial agents, 43-45
  blood levels, 44
  therapy, 313-328
    in septicaemia, 240
Antiseptics, 51-53
Antitetanus immunoglobulin, 230
Antituberculous chemotherapy, 322
Anthracoid bacilli, 114
Anthrax, 296
API system, 41
Arthritis infective, 260
Arthropod-borne infection, 154
Aschoff nodule, 163
Aspiration pneumonia, 176

Autoclaves, 48-50
Autotrophs, 15
Auxotrophs, 17

β-lactamase, 60, 314
  from gonococcus, 110
B-lymphocytes, 146
Bacillus, 113
  anthracis in anthrax, 296
  cereus food poisoning, 185
Bacillary dysentery, 190-192
Bacteraemia, 238
Bacteraemic shock, 240
Bacitracin sensitivity test, 63
Bacteria of medical importance, 55 *et seq.*
see also various organisms
Bacterial biology, 1 *et seq.*
  chromosome, 19
  culture, 35-40
  disease, 137 *et seq.*
    treatment and prevention, 311 *et seq.*
  enzymes, 18
  epidemics, 156
  genetics, 22-32 *see also* Genetics
    bacterial
  growth, 14-21 *see also* Growth of
    bacteria
    division, 19
  identification, 40-43
  infection and PUO, 235
  meningitis, 211
  structure, 6-11 *see also* Structure of
    bacteria
  taxonomy, 11-13
  toxins, 149
  typing, 42
  zoonoses, 291
Bactericidal drugs, 313
Bacteriocin typing, 43
  of *Proteus* spp., 77
    Pseudomonas spp., 81
Bacteriophage, 28
  typing, 43
    of salmonellae, 76
    Staph. aureus,, 59
    V. cholerae, 87
Bacteriuria, 205
Bacteroides, 120
Base substitution in mutation, 26
BCG vaccine, 254, 334
  in leprosy, 257
Bedpans, 51
Beneckea, 88
Bifidobacterium, 122
Binary fission, 19
Biochemical tests in identification, 41

Bi-polar staining, 78
Black Death, 298
Blepharitis, 271
Blood culture, 40, 159
  in endocarditis, 244
    enteric fever, 201
  septicaemia, 240
Boilers, 51
Boils, 218
Bone infections, 258
  tuberculous, 248
Bordet-Gengou medium, 95
Bordetella, 95
Borrelia, 129
Botulism, 119, 198
Bovine tuberculosis, 251
Branhamella catarrhalis, 112
Bronchitis, 167, 168-171
  treatment, 169, 170
  vaccines, 170
Bronchopneumonia, 173
  tuberculous, 247
Brucella, 93, 292
  antibodies, 294
Brucellin test, 295
Brucellosis, 292-296
  in animals, 295
Bubonic plague, 298
Burns infection of, 225

Campylobacter, 88
Cancrum oris, 287
Candida, 133-135
  albicans in candidosis, 165
Candidosis, 165
  genital, 267
  oral, 284, 285
Capsules, 8
  of *Haemophilus*, 91
  swelling of pneumococci, 70
Carbon dioxide, 15, 18
  metabolism, 15
Carbuncles, 218
Cardiac disease and infective
    endocarditis, 242
Cardiovascular syphilis, 268
Cariogenic bacteria, 278
Carriers, 152
  in epidemics, 155
    spread of cholera, 193
    enteric fever, 200-203
Catheterization and infection, 210, 306
Cell division, 19
  mediated immunity, 147, 329
    in tuberculosis, 250
  wall, 8-10

Cellulitis, 220
  and dental disease, 282
  orbital, 273
Central sterile supply, 50
Cephalosporins, 315
Cerebrospinal fluid, 159
  laboratory examination, 214
Cervicitis, non-specific, 265
Cetrimide agar, 80
Chemotherapy, antituberculous, 322
Chest infections, diagnosis, 176
Childhood fevers in general practice, 309
Chinese-character arrangement, 97
Chlamydial infections, diagnosis, 265, 273
  conjunctivitis, 273
  genital infection, 265
Chocolate agar, 109
Cholera, 85-87, 192-194
  vaccine, 338
Chorioretinitis, 274
Choroiditis, 274
Citrobacter, 78
Classification of bacteria, 11-13
Clindamycin, 320
Clostridium, 115-119
  botulinum, 119, 198
  difficile, 119
  tetani, 118
  welchii, 116
  wound infection, 226-230
Coagulase as aggressin, 149
  in identification, 42
  test, 58
Colicin typing, 75
Coliform bacilli, 71-79
Colitis, antibiotic-associated, 194-196
Colon bacterial flora, 140
Colonial morphology, 40
Commensal bacteria, 139
Complement, 146
  fixation, 46
Conjugation, 29
Conjunctivitis, 262, 271, 272
Consumption, 247
Contact spread of infection, 153
Coombs' test, 45
Corynebacterium, 97-101
  diphtheriae, 97-101
    in diphtheria, 164
    toxin, 99
  other species, 101
Cotrimoxazole, 316
Cough plate, 172
Croup, 168
Cryptogenic abscesses, 232
CSF, 159, 214
Culture of bacteria, 35-40

Cystic fibrosis, 171
Cystitis, 205
Cytoplasmic membrane, 10

Dark-ground microscopy, 34
Defence mechanisms of host, 143-148
  immune system, 144-148
  non-specific, 144
Dehydration in cholera, 193
Delayed hypersensitivity, 147
  in brucellosis, 295
    leprosy, 256
    tuberculosis, 250
Deletion mutation, 26
Dental infections, 275-290
  calculus, 281
  caries, 275-282
  non-specific localised, 275-284
  plaque, 275-277
  prevention, 279
  specific localised, 284-288
Denture stomatitis, 285
Diarrhoeal diseases, 179-197
  antibiotic-associated, 194
  causal organisms, 180
  cholera, 192-194
  dysentery, 190-192
  infantile, 188
  investigation, 179
  non-bacterial infective, 196
  travellers', 188, 189
  yersiniosis, 196
Dick test, 65
Diene's phenomenon, 77
  stain, 131
Dip slide, 158
Diphtheria, 101, 164
  toxin, 99
  toxoid, 340
Disease bacterial, 137 *et seq.*
Disc diffusion tests, 43
Disinfectants, 51-53
  hospital use, 53
DNAases streptococcal, 64
Draughtsman colonies, 69
Droplet spread of infection, 153
Drugs antimicrobial, 313-328
  combinations, 326
  resistance, 323
  therapeutic principles, 324-328
  therapy in tuberculosis, 253-254
  toxicity, 325
Dry heat sterilisation, 50
Dust in spread of infection, 153
Dysentery, 75, 190-192

Ear flora, 142
Earache, 166
El Tor *V. cholerae*, 86
Electron microscopy, 34
Elek test, 99
Empyema, 176
Endocarditis infective, 241-246
  diagnosis, 244
  prosthetic valve in, 244
  treatment, 245
Endogenous infection, 151
Endotoxic shock, 240
Endotoxins, 73, 149
  assay, 240
Enteric fever, 76, 200-204
  diagnosis, 201
  treatment, 203
Enterobacter, 78
Enterobacteria, 71-79
  antibiotic sensitivity, 73
  laboratory characteristics, 72
Enterococci, 63, 67
Enterotoxin, 118
  demonstration, 182
Enzymes bacterial, 18
  in identification, 42
Epidemiology, 151-157
  measurements, 155
  surveillance, 156
Epiglottitis, 168
Erysipelas, 219
Erysipeloid, 302
Erysipelothrix 102
Erythrogenic toxin, 65
Erythromycin, 320
Escherichia coli, 71, 74
  in diarrhoea, 188, 190
  urinary infections, 206
Ethambutol, 322
Eubacterium, 123
Eukaryotic cells, 6
Exogenous infection, 151
Exotoxins, 149
  of enterobacteria, 73
  V. cholerae, 193
Eye infections, 271-274
  diagnosis, 273

Faecal flora in wound sepsis, 224
Faecal-oral infection, 153
  in cholera, 193
  in dysentery, 192
Faeces, bacteria in, 140
  culture, 201
  safe disposal, 51
  specimens, 158
  in food-poisoning, 181
Fermentation of sugars, 15

F factors, 29-31
Flagella, 7
Fleming, 4
Flora bacterial, 139-142
Fluorescence microscopy, 34
Follicular tonsillitis, 161
Fomites, 153
Food as source of infection, 152
  hygiene, 182, 183, 185, 186
  poisoning due to *B. cereus*, 114, 185
    Campylobacter, 186
    Clostridium welchii, 118, 184
    salmonellae, 76, 180-182
    staphylococci, 182
    Vibrio parahaemolyticus, 187
Fragilis group of *Bacteroides*, 120, 121
Francisella, 96
Fusobacterium, 120

Gas gangrene, 118, 227
  in cellulitis, 222
Gastrointestinal flora, 140
Gene transfer, 26-31
Generation time, 20
Genetics bacterial, 22-32
  DNA, 23
  gene transfer, 26-31
  mutation, 25
  transposition, 31
Genital candidosis, 267
  tract flora, 141
Genitourinary tuberculosis, 248
Genome, 6, 19, 23
General practice infection incidence, 308
Gingivitis prevention, 280
  ulcerative, 286
Gingivostromatitis, 165
Glomerulonephritis, 163
Gonococcal conjunctivitis, 262, 272
  vulvovaginitis, 263
Gonococcus, 109 *see also N. gonorrhoeae*
Gonorrhoea, 110, 262-265
  diagnosis, 263
  oropharyngeal, 289
Gram stain and cell wall structure, 6, 8
  in classification, 12
  method, 34
Griffith types, 65
Growth of bacteria, 14-21
  carbon metabolism, 15
  environmental conditions, 18
  growth cycle, 19
    factors, 17
      for *Haemophilus*, 90
  in vivo, 21
  nitrogen metabolism, 17
  nutrient limitation, 20, 21
    uptake, 17

Haemolysis by streptococci, 62, 65, 68
Haemophilus, 90-93
  ducreyi, 92
  influenzae, 91
    in bronchitis, 169
      meningitis, 213
      pneumonia, 174
    pathogenicity, 91
Heaf test, 251
Heterotrophs, 15
Herd immunity, 154
Hospital wound infection, 222-226
Host-parasite relationship, 143-150
Humoral response, 146
Hyaluronidase, 64, 149
Hypernatraemia, 188
Hypersensitivity delayed, 147 *see also*
  Delayed hypersensitivity

Identification of bacteria, 40-43
Ig, 146, 147
Immune system, 146-148
Immunisation prophylactic, 329-343
  active, 331-341
    against tetanus, 230
      tuberculosis, 254
    and epidemics, 157
    complications, 334
    official policy, 333, 341
    passive, 341-343
Immunity, 329
Immunodeficiency and infection, 304
Immunofluorescence, 34, 46
Immunoglobulins, 146, 147, 341
Immunology in diagnosis, 45
Immunosuppression and disease, 144,
  305
Impetigo, 219
Incidence rate, 156
Inclusion conjunctivitis, 272
Inclusions, bacterial, 11
Incubation, cultures, 40
  period, 156
Infantile gastroenteritis, 188
Infection, in compromised patients,
  304-307
  in general practice, 308
  initiation, 143
  of bone, 258
    joints, 260
    respiratory tract, 161-178
    urinary tract, 205-211
  routes, 153
  sources, 151-153
  subclinical, 152
Infectious mononucleosis, 165

Infective arthritis, 260
  endocarditis, 241 *see also* Endocarditis
Influenza in general practice, 309
  vaccine, 340
Ingestion, infection by, 153
Inhibition zones, 43
Inoculation method, 36
Insertion mutation, 26
Invasiveness, 148
Isolation in laboratory diagnosis, 158
Isoniazid, 322
Intensive care units and infection, 305,
  306
Intestinal flora, 140

Jenner, 4
Joint infections, 260
  prosthetic joints, 261
  tuberculous, 248

Kauffmann-White scheme, 76
Keratitis, 272
Kernig's sign, 211
Klebicin bacteriocin typing, 77
Klebsiella, 71, 76
Koch, 4

Laboratory diagnosis, 158
  examinations in PUO, 236
  methods, 33-46
    bacterial identification, 40-43
    culture, 35-40
    incubation, 40
    serology, 45
    tests of antimicrobials, 43-45
Lactobacillus 124
  in caries, 278
Lancefield grouping, 62, 63
  group A, 64-66
  group B, 66
  group C, 66
  group D, 67
  group G, 67
Laryngitis, 167, 168
Lecithinase, 117
  in identification, 42
Legionella, 125
Legionnaire's disease, 126, 173, 175
Leptospirosis, 300
Lepromin test, 256
Leptospira, 130
Leprosy, 104, 254-257
L-forms, 9
Light microscopy, 33
Lincomycin, 320

Lister, 4
Listeria, 102
Listeriosis, 302
Lobar pneumonia, 173
Lockjaw, 228
Loeffler's serum medium, 97
Lowenstein-Jensen medium, 104
Lung abscess, 176
  infections, 172
Lymphadenitis staphylococcal, 287
  tuberculous, 288
Lymphangitis, 149
Lymphocytes, 146, 147
Lymphokines, 148
Lysed blood agar, 109

Macrophages, 145
Madura foot, 108
Malignant pustule, 296
Mannitol fermentation, 59
Mantoux test, 251
Maternal antibody, 333
McFadyean's reaction, 113
Measles vaccine, 336
Media culture, 36-39
  transport, 40
Medically important bacteria, 55 *et seq.*
Melaninogenicus/oralis group, 120, 121
Meningism, 212
Meningitis, 211-216
  antibiotic treatment, 214
  causal organisms, 212
  clinical features, 211
  diagnosis, 214
  in leptospirosis, 300
  meningococcal, 111
  pathogenesis, 213
  tuberculous, 248
Meningococcus, 109 *see also Neisseria meningitidis*
Mesosomes, 11
Metabolism of bacteria, 14-21
Metachromatic granules, 97
Metronidazole, 319
Micrococci, 60
Microscopy, 33-35
Middle ear infections, 166
Miliary tuberculosis, 248
Milk in spread of brucellosis, 293
  pasteurisation, 53
Minimum inhibitory concentration, 44
Moraxella, 83
Mortality rate, 156
Motility in spirochaetes, 128
Mumps vaccine, 336
Mutation, 25
Mycetoma, 108

Mycobacterium atypical spp., 105
  bovis, 104
  leprae, 104
    in leprosy, 255
  tuberculosis, 103
    culture from specimens, 252, 253
    in meningitis, 213
    tuberculosis, 247
Mycoplasma, 9, 131
  in bronchitis, 169
    primary atypical pneumonia, 174
Myocarditis streptococcal, 162

Nagler reaction, 117
Neisseria commensal, 111
  gonorrhoeae, 109
    in gonorrhoea, 262
  meningitidis, 110
    in meningitis, 212
Neonatal meningitis, 212
  necrotizing enterocolitis, 196
  skin sepsis, 219
Neurosyphilis, 268
Neutrophil polymorphonuclear
  leucocytes, 145
Nitrogen metabolism, 17
Nocardia, 107
Non-specific defence mechanisms, 144
  genital infections, 265
Normal flora, 139-142
Nose bacterial flora, 140
Nuclear material of bacteria, 11
Nutrition of bacteria, 14-21

O antigens, 149
Ophthalmia neonatorum, 262, 272
Oral infections, 275-290
  candidosis, 284, 285
  syphilis, 288
  tuberculosis, 288
Orbital cellulitis, 273
Oropharyngeal flora, 140
  gonorrhoea, 289
Orthopaedics, infection in, 225
Osteomyelitis, 258-259
  and dental disease, 283
Otitis media, 166, 167
Oxidase test, 109, 263
Oxidation of sugars, 15
Oxygen in bacterial growth, 18

Panophthalmitis, 273
Paratyphoid fever, 200-204
Parvobacteria, 90-96

Passive immunisation, 341-343
  immunity, 330
Pasteur, 4
Pasteurella, 95, 300
Pasteurisation, 53
Pathogenic synergy, 122
Pathogenicity, 148 *see also* various
    organisms
Pemphigus neonatorum, 219
Penicillins, 314
  hypersensitivity to, 326
  resistance to, 60
Penicillium notatum, 313
Peptococcus, 123
Peptone, 38
Periapical abscess, 282
Periodontal disease, 275-282
  prevention, 280
Peritonitis, 226
Pertussis, 171
  cough plate, 172
  vaccine, 157, 338
Phage conversion, 29
  typing, 43 *see also* Bacteriophage
      typing
Phagocytosis, 145
Pharyngeal infections, 161-166
  diagnosis, 166
Phase contrast microscopy, 34
Phenotypic variants, 22
Phosphatase test for *Staph. aureus*, 59
Phospholipase C, 117, 149
Physiological adaptation, 22
Pittman types, 91
Pili, 7
Pinta, 129
Plague, 78, 298
  vaccine, 338
Plaque dental, 275
  and gingivitis, 280
Plasmids, 29-32
Plesiomonas, 83
Pneumococcus, 69 *see also Strep.
    pneumoniae*
  vaccine, 339
Pneumonia, 172-176
  causal agents, 173
  treatment, 175
Pneumonic plague, 298
Poliomyelitis vaccine, 335
Polymorphs, 145
Precipitation, 45
Prevention of bacterial disease, 311 *et
    seq.*
Primary atypical pneumonia, 173
  serological tests, 178
Prokaryotic cells, 6
Propionibacterium, 102
Prophylactic immunisation, 329-343

Prostheses and infection, 307
  heart valves, 244
  joints, 261
Proteolytic clostridia, 116
Proteus, 71, 77
Protoplasts, 8
Prototrophs, 17
Providencia, 78
Pseudomembranous colitis, 119
Pseudomonas, 80-82
Puerperal sepsis, 225
Pulmonary tuberculosis, 247
PUO, 234-237
Pure culture isolation, 37
Pyelonephritis, 205
  in children, 209
Pyrexia of unknown origin, 234-237
Pyocin typing, 81
Pyogenic streptococci, 62-68

Quellung reaction, 70
Quinsy throat, 162

Rat bite fever, 303
Recombination genetic, 27
Reiter's disease, 265
Relapsing fevers, 129
Resistance to antibiotics, 323
Respiratory tract flora, 140
  infections, 161-178
    bronchi, 167
    diagnosis, 176
    in general practice, 308
    lungs, 172
    middle ear, 166
    pharynx, 161-166
    sinuses, 166
    throat, 161-166
    trachea, 167
    tuberculosis, 247
Reticulo-endothelial cells, 145
R factors, 31
Rheumatic fever, 162
Ribosomes, 11
Rice-water stools, 193
Rifampicin, 322
Risus sardonicus, 228
Rubella vaccine, 337

Sabouraud's medium, 133
Saccharolytic clostridia, 116
Saliva bacterial flora, 140
Salmonella, 71, 75, 76
  in food-poisoning, 180-182
  paratyphoid, 201
  typhoid, 200

Scalded skin syndrome, 219
Scarlet fever, 162
Schick test, 100
Schuffner test, 301
Selective media, 37
  for Bord. pertussis, 95
    Candida, 133
    C. diphtheriae, 97
    enterobacteria, 72
    mycobacteria, 104
    Neisseria, 109
    Pseudomonas, 80
    Staphylococcus, 58
    Streptococcus, 63
    V. cholerae, 85-87
Selective pressure, 22
Sensitivity tests, 43-45
Sepsis, 217-233
  abscesses, 230
  cellulitis, 220
  peritonitis, 226
  skin infection, 217-220
  wound infection, 222-226
    clostridial, 226-230
Septic abortion, 225
  shock, 240
Septicaemia, 149, 238-241
  causal organisms, 238
  treatment, 240
Serology, 45
  in diagnosis, 159
    syphilis, 268
Serous fluid specimens, 159
Serratia, 78
Sex pilus, 29
Sexually transmitted diseases, 154,
    262-270
Shigella, 71, 74
  in dysentery, 190
Shock septic, 240
Sialadenitis, 283
Sinusitis, 166, 167
Skin flora, 142
  infection, 217-220
    diagnosis, 220
  test in leprosy, 256
    tuberculosis, 250
Smallpox vaccine, 337
Smoker's cough, 168
Sonne dysentery, 191
Soil as source of infection, 152
Specimens, bacteriological, 158
  collection, 158
  serology, 159
  transport, 159
Spheroplasts, 9
Spirochaetes, 127-130

Spores, 11, 47
  of Bacillus, 113, 114
    Clostridium, 115
Sputum laboratory examination, 177
  specimens, 158
Staining, 33, 34
Staphylococcus, 57-61
  antibiotic sensitivity, 60
  pathogenicity, 60
    food-poisoning, 182
    lymphadenitis, 287
    sticky eye, 272
  species, 57-60
Sterilisation and disinfection, 47-54
  centralised facilities, 50
  disinfectants and antiseptics, 51
    hospital use, 53
  methods, 48
Sticky eye, 272
Stormy clot, 117
Streptobacillus, 84
Streptococcal sore throat, 161
  complications, 162
Streptococcus, 62-70
  classification, 62
  Lancefield group A, 64-66
    B, C, D and G, 66, 67
  mutans in caries, 277
  pneumoniae, 69
    in bronchitis, 169
      meningitis
      pneumonia, 173
  pyogenes, 62-68
    antibiotic sensitivity, 66
    laboratory characteristics, 64
    pathogenicity, 65
  viridans, 68
Streptokinase, 64
Streptolysins, 65
Structure of bacteria, 6-11
  cell wall, 8-10
  cytoplasmic membrane, 10
  external structures, 7
  internal structures, 7, 11
Styes, 218, 271
Subclinical infections, 152
Sulphonamides, 316-318
Sulphur granules, 106, 284
Surgical sepsis, 222-226
Swabs, specimens, 159
Swarming of clostridia, 116
Sycosis barbae, 218
Synergy in drug therapy, 326
  pathogenic, 122
Syphilis, 128, 267
  diagnosis, 268
  oral, 288

TAB vaccine, 337
Taxonomy, bacterial, 11-13
Teichoic acid, 8
Tests in identification, 41
Tetanus, 119, 228
  neonatorum, 229
  toxins, 118, 228
  toxoid, 230, 341
Tetracyclines, 319
Thayer-Martin medium, 109
Therapy antimicrobial, 313-328
Throat infections, 161-166
  diagnosis, 166
  swabs, 166
Thrombi in endocarditis, 243
Thrush, 135, 165, 285
  vaginal, 267
Titre, 45
T-lymphocytes, 147
Tonsillitis, acute follicular, 161
  exudative, 165
  ulcerative, 165
Toxic epidermal necrolysis, 219
Toxicity of drugs, 325
Toxins bacterial, 149
  of *Clostridium* spp., 116-118
    C. diphtheriae, 99
    Staph. aureus, 59
    Strep. pyogenes, 64, 65
    V. cholerae, 87
Toxoids, 149, 331, 340
Tracheitis, 167
Tracheostomies and infection, 225
Transduction, 28
Transformation, 27
Transplacental infection, 154
Transport of specimens, 40, 159
Transposition, 31
Travellers' diarrhoea, 188, 189
Treatment of bacterial disease, 311 *et seq.*
Treponema pallidum, 128
  in syphilis, 267
Trichomoniasis, 266
Trimethoprim, 316-318
Triple vaccine, 230, 341
Tuberculin test, 250
Tuberculosis, 104, 247-254
  control, 254
  delayed-type hypersensitivity, 250
  diagnosis, 252
    differential, 176
  epidemiology, 251
  oral, 288
  renal, 208
  treatment, 253
Tuberculous bronchopneumonia, 247
  lymphadenitis, 288
  meningitis, 248

Tularaemia, 300
Typing of bacteria, 42
Typhoid fever, 200-204
  vaccine, 337

Ulcerative gingivitis, 286
Undulant fever, 292
Ureaplasma, 131, 132
Urethral flora, 141
  syndrome, 206
Urethritis, 205
  non-specific, 265
Urinary antimicrobial drugs, 74
  tract infections, 205-211
    causal organisms, 206
    diagnosis, 207
    Proteus in, 77
    Pseudomonas in, 81
    treatment, 209
    surgical, 210
Urine specimens, 158, 207
  semi-quantitative culture, 207, 208

Vaccination *see* Immunisation
Vaccines, 331
  administration, 333
  combined, 333
  currently in use, 334-341
  in bronchitis, 170
  preparation, 332
  schedule for U.K., 341
  triple, 341
Vaginal flora, 141
  barrier to infection, 145
Vaginitis due to *H. vaginalis*, 269
Van Leeuwenhook, 3
Variation bacterial, 22-25
Vector-borne infection, 154
Vegetations in endocarditis, 243
Veillonella, 123
Vibrio cholerae, 85-87
  cholerae in cholera, 192
  parahaemolyticus in food-poisoning, 187
Vincent's angina, 129, 165, 287
  gingivitis, 286
Viral pneumonia, 175
  throat infections, 161
Viridans streptococci, 63, 68
Virulence, 148
Volutin granules, 97

Water-borne infection, 152
  cholera, 194
  dysentery, 192
  enteric fever, 204
  Public Health aspects, 54

Wassermann reaction, 268
Weil's disease, 300
Wet films, 33, 34
Whooping cough, 95, 171
  vaccine, 157, 338
Woolsorter's disease, 296
Widal test, 202
Wound infection, 222-226
  clostridial, 226-230
  diagnosis, 225

Yaws, 129
Yeast-like fungi, 133-135
Yersinia, 71, 78
  in plague, 298

Yersiniosis, 196

Ziehl-Neelsen stain, 35, 103
  diagnosis of tuberculosis, 252
Zones of inhibition, 43
Zoonoses, 291-303
  anthrax, 296
  brucellosis, 292-296
  leptospirosis, 300
  listeriosis, 302
  pasteurella, 300
  plague, 298
  tularaemia, 300